Artificial Intelligence and Digital Systems Engineering

Analytics and Control Series

Series Editor:
Adedeji B. Badiru
Air Force Institute of Technology, Dayton, Ohio, USA

Decisions in business, industry, government, and the military are predicated on performing data analytics to generate effective and relevant decisions, which will inform appropriate control actions. The purpose of the focus series is to generate a collection of short-form books focused on analytic tools and techniques for decision making and related control actions.

Mechanics of Project Management
Nuts and Bolts of Project Execution
Adedeji B. Badiru, S. Abidemi Badiru, and I. Adetokunboh Badiru

The Story of Industrial Engineering
The Rise from Shop-Floor Management to Modern Digital Engineering
Adedeji B. Badiru

Innovation
A Systems Approach
Adedeji B. Badiru

Project Management Essentials
Analytics for Control
Adedeji B. Badiru

Sustainability
Systems Engineering Approach to the Global Grand Challenge
Adedeji B. Badiru and Tina Agustiady

Operational Excellence in the New Digital Era
Adedeji B. Badiru and Lauralee Cromarty

Artificial Intelligence and Digital Systems Engineering
Adedeji Badiru

For more information on this series, please visit: https://www.routledge.com/Analytics-and-Control/book-series/CRCAC

Artificial Intelligence and Digital Systems Engineering

Adedeji B. Badiru

CRC Press is an imprint of the
Taylor & Francis Group, an **informa** business

First edition published 2022
by CRC Press
6000 Broken Sound Parkway NW, Suite 300, Boca Raton, FL 33487-2742

and by CRC Press
2 Park Square, Milton Park, Abingdon, Oxon, OX14 4RN

CRC Press is an imprint of Taylor & Francis Group, LLC

Library of Congress Cataloging-in-Publication Data
Names: Badiru, Adedeji Bodunde, 1952- author.
Title: Artificial intelligence and digital systems engineering / Adedeji Badiru.
Description: First edition. | Boca Raton, FL: CRC Press, 2022. |
Series: Analytics and control |
Includes bibliographical references and index.
Identifiers: LCCN 2021023373 (print) | LCCN 2021023374 (ebook) |
ISBN 9780367545475 (hardback) | ISBN 9780367545482 (paperback) |
ISBN 9781003089643 (ebook)
Subjects: LCSH: Systems engineering—Data processing. | Artificial intelligence.
Classification: LCC TA168 .B238 2022 (print) |
LCC TA168 (ebook) | DDC 620.001/171—dc23
LC record available at https://lccn.loc.gov/2021023373
LC ebook record available at https://lccn.loc.gov/2021023374

ISBN: 978-0-367-54547-5 (hbk)
ISBN: 978-0-367-54548-2 (pbk)
ISBN: 978-1-003-08964-3 (ebk)

DOI: 10.1201/9781003089643

Typeset in Times
by codeMantra

Dedicated to the embrace of intelligence, both natural and artificial

Contents

Preface xi
Acknowledgments xiii
About the Author xv

1 Understanding AI **1**
Introduction 1
Historical Background 2
Origin of Artificial Intelligence 2
Human Intelligence versus Machine Intelligence 4
Natural Language Dichotomies 6
The First Conference on Artificial Intelligence 8
Evolution of Smart Programs 9
Branches of Artificial Intelligence 12
Neural Networks 13
Emergence of Expert Systems 15
Conclusion 17
References 18

2 Expert Systems: The Software Side of AI **19**
Expert Systems Process 19
Expert Systems Characteristics 19
Expert Systems Structure 21
The Need for Expert Systems 23
Benefits of Expert Systems 24
Transition from Data Processing to Knowledge Processing 25
Heuristic Reasoning 25
User Interface 26
Application Roadmap 27
Symbolic Processing 27
Future Directions for Expert Systems 29
Academia–Industry Cooperation for Expert Systems 30
Sample of Expert Systems Applications 33

3 Digital Systems Framework for AI **43**
Digital Framework for AI 43
Digital Engineering and Systems Engineering 44
Introduction to DEJI Systems Model 45
Application of DEJI Systems Model to Systems Quality 47
The Waterfall Model 47
The V-Model 48
Spiral Model 49
Defense Acquisitions University SE Model 50
Walking Skeleton Model 50
Object-Oriented Analysis and Design (OOAD) 52
Digital Data Input–Process–Output 53
Digital Collaborations 57
Lean and Six Sigma in AI 60
Humans in the Loop in AI Implementations 61
Conclusion 62
References 63

4 Neural Networks for Artificial Intelligence **65**
Introduction 65
Definition of a Neurode 67
Variations of a Neurode 68
Single Neurode: The McCullough-Pitts Neurode 69
Single Neurode as Binary Classifier 70
Single Neurode Perceptron 71
Single-Layer Feedforward Network: Multi-Category SLP 71
Associative Memory 71
Correlation Matrix Memory 72
Widrow–Hoff Approach 73
LMS Approach 73
Adaptive Correlation Matrix Memory 74
Error-Correcting Pseudo-Inverse Method 74
Self-Organizing Networks 74
Principal Components 75
Clustering by Hebbian Learning 76
Clustering by Oja's Normalization 77
Competitive Learning Network 78
Multiple-Layer Feedforward Network 78
Multiple-Layer Perceptron 78
XOR Example 79
Back-Error Propagation 79

Variations in the Back-Error Propagation Algorithm 80
Learning Rate and Momentum 81
Other Back-Error Propagation Issues 82
Counter-Propagation Network 83
Radial Basis Networks 84
Interpolation 84
Radial Basis Network 85
Single-Layer Feedback Network 87
Discrete Single-Layer Feedback Network 88
Bidirectional Associative Memory 89
Hopfield Network 90
Summary 92
References 93

5 Neural-Fuzzy Networks for Artificial Intelligence 95
Technology Comparisons 95
Neurons Performing Fuzzy Operations 98
Neurons Emulating Fuzzy Operations 98
Neural Network Performing Fuzzy Inference 100
Regular Neural Network with Crisp Input and Output 100
Regular Neural Network with Fuzzy Input and Output 101
Fuzzy Inference Network 102
Adaptive Neuro-Fuzzy Inference System (ANFIS) 103
Commutative Applications 104
Clustering and Classification 105
Classification 106
Multilayer Fuzzy Perceptron 107
References 108

Index 111

Preface

This book presents a short-form systems introduction to artificial intelligence (AI) and digital systems engineering. It is not intended to be a comprehensive treatment of AI or digital systems engineering. Rather, it presents a quick view for general readership to understand more of what AI portends for the society in the present digital age. The book uses a systems-based approach to show how AI is pervasive in all facets of endeavors, including business, industry, government, the military, and even academia. The systems approach facilitates process design, evaluation, justification, and integration. A key power of the book is that it explicitly highlights the role of integration in any AI implementation of objectives. AI is the modern digital lifeline of operational excellence. This book presents brief synopsis on techniques and methodologies for achieving the transfer of science and technology assets for AI applications. AI can mean different things to different people. This book presents conceptual and operational definitions of AI. Emphasis is placed on the context related to the theme of systems thinking, how each element plays into the overall structure of an AI infrastructure. Some definitions are used within the framework of conceptual processes, while some are used within the platform of technology. Some organizations can be innovative in utilizing AI tools, while some organizations are adept in researching and developing new AI tools. For this reason, a clarification of the various roles, meanings, and contexts of AI is essential. This book meets that purpose, in a focused book format. Foremost in this process of AI application is the role of organizational leadership in embracing AI applications and its ramifications. This book explains the differences.

Acknowledgments

I acknowledge and appreciate my various colleagues, students, and professional affiliates, who have, over the years, inspired me to continue writing about the topic that means a lot to me within the realm of teaching, research, and intellectual outreach. Too many to name individually, I extend my profound salutation to all.

About the Author

Adedeji Badiru is a professor of systems engineering at the Graduate School of Engineering and Management at the Air Force Institute of Technology. He was previously a professor and department head of Industrial Engineering at the University of Tennessee and professor of Industrial Engineering at the University of Oklahoma. He is a registered professional engineer (PE), a certified project management professional (PMP), and a fellow of the Institute of Industrial & Systems Engineers. He is the author of over 30 books, 34 book chapters, and 80 technical journal articles. He is a member of several professional associations and scholastic honor societies.

Understanding AI

1

INTRODUCTION

Artificial intelligence (AI) is not just one single thing. It is a conglomerate of various elements, involving software, hardware, data platform, policy, procedures, specifications, rules, and people intuition. How we leverage such a multifaceted system to do seemingly intelligent things, typical of how humans think and work, is a matter of systems implementation. This is why the premise of this book centers on a systems methodology. In spite of the recent boost in the visibility and hype of artificial intelligence, it has actually been around and toyed with for decades. What has brought AI more to the forefront nowadays is the availability and prevalence of high-powered computing tools that have enabled the data-intensive processing required by AI systems. The resurgence of AI has been driven by the following developments:

- Emergence of new computational techniques and more powerful computers
- Machine learning techniques
- Autonomous systems
- New/innovative applications
- Specialized techniques: Intelligent Computational Search Technique Using Cantor Set Sectioning
- Human-in-the-loop requirements
- Systems integration aspects

As long ago as the mid-1980s, the author has led many research and development projects that embedded AI software and hardware into conventional human decision processes. AI has revolutionized and will continue to revolutionize many things we see and use around us. So, we need to pay attention to the emerging developments.

DOI: 10.1201/9781003089643-1

HISTORICAL BACKGROUND

The background of Al has been characterized by controversial opinions and diverse approaches. The controversies have ranged from the basic definition of intelligence to questions about the moral and ethical aspects of pursuing AI. However, despite the unsettled controversies, the technology continues to generate practical results. With increasing efforts in AI research, many of the prevailing arguments are being resolved with proven technical approaches. Expert system, the main subject of this book, is the most promising branch of AI.

"Artificial intelligence" is a controversial name for a technology that promises much potential for improving human productivity. The phrase seems to challenge human pride in being the sole creation capable of possessing real intelligence. All kinds of anecdotal jokes about AI have been offered by casual observers. A speaker once recounted his wife's response when he told her that he was venturing into the new technology of AI. "Thank God, you are finally realizing how dumb I have been saying you were all these years," was alleged to have been the wife's words of encouragement. One whimsical definition of AI refers to it as the "Artificial Insemination of knowledge into a machine." Despite the deriding remarks, serious embracers of AI may yet have the last laugh. It is being shown again and again that AI may hold the key to improving operational effectiveness in many areas of application. Some observers have suggested changing the term "Artificial Intelligence" to a less controversial one such as "Intelligent Applications (IA)." This refers more to the way that computers and software are used innovatively to solve complex decision problems.

Natural Intelligence involves the capability of humans to acquire knowledge, reason with the knowledge, and use it to solve problems effectively. It also refers to the ability to develop new knowledge based on existing knowledge. By contrast, *Artificial Intelligence* is defined as the ability of a machine to use simulated knowledge in solving problems.

ORIGIN OF ARTIFICIAL INTELLIGENCE

The definition of intelligence had been sought by many ancient philosophers and mathematicians including Aristotle, Plato, Copernicus, and Galileo. These great philosophers attempted to explain the process of thought and understanding. The real key that started the quest for the simulation of intelligence did not occur, however, until the English philosopher Thomas Hobbes put forth an interesting concept in the 1650s. Hobbes believed that thinking consists of symbolic operations and that everything in life can be represented mathematically. These beliefs directly led to the notion that a machine capable of carrying out mathematical operations on symbols could imitate human thinking. This is the basic driving force behind the AI effort. For that reason, Hobbes is sometimes referred to as the grandfather of AI.

While the term "Artificial Intelligence" was coined by John McCarthy in 1956, the idea had been considered centuries before. As early as 1637, Rene Descartes was conceptually exploring the ability of a machine to have intelligence when he said:

For we can well imagine a machine so made that it utters words and even, in a few cases, words pertaining specifically to some actions that affect it physically. However, no such machine could ever arrange its words in various different ways so as to respond to the sense of whatever is said in its presence—as even the dullest people can do.

Descartes believed that the mind and the physical world are on parallel planes that cannot be equated. They are of different substances following entirely different rules and can, thus, not be successfully compared. The physical world (i.e., machines) cannot imitate the mind because there is no common reference point.

Hobbes proposed the idea that thinking could be reduced to mathematical operations. On the other hand, Descartes had the insight into functions that machines might someday be able to perform. But he had reservations about the concept that thinking could be simply a mathematical process.

The 1800s was an era that saw some advancement in the conceptualization of the computer. Charles Babbage, a British mathematician, laid the foundation for the construction of the computer, a machine defined as being capable of performing mathematical computations. In 1833, Babbage introduced an Analytical Engine. This computational machine incorporated two unprecedented ideas that were to become crucial elements in the modern computer. First, it had operations that were fully programmable, and second, the engine could contain conditional branches. Without these two abilities, the power of today's computers would be inconceivable. Babbage was never able to realize his dream of building the analytic engine due to a lack of financial support. However, his dream was revived through the efforts of later researchers. Babbage's basic concepts could be observed in the way that most computers operate today.

Another British mathematician, George Boole, worked on issues that were to become equally important. Boole formulated the "laws of thought" that set up rules of logic for representing thought. The rules contained only two-valued variables. By this, any variable in a logical operation could be in one of only two states: yes or no, true or false, all or nothing, 0 or 1, on or off, and so on. This was the birth of digital logic, a key component of the AI effort.

In the early 1900s, Alfred North Whitehead and Bertrand Russell extended Boole's logic to include mathematical operations. This not only led to the formulation of digital computers but also made possible one of the first ties between computers and thought process.

However, there was still a lack of an acceptable way to construct such a computer. In 1938, Claude Shannon published "A Symbolic Analysis of Relay and Switching Circuits." This work demonstrated that Boolean logic consisting of only two-variable states (e.g., on–off switching of circuits) can be used to perform logic operations. Based on this premise, the ENIAC (Electronic Numerical Integrator and Computer) was built in 1946 at the University of Pennsylvania. The ENIAC was a large-scale fully operational electronic computer that signaled the beginning of the first generation of computers. It could perform calculations 1,000 times faster than its electromechanical predecessors. It weighed 30 tons, stood two stories high, and occupied 1,500 square feet of floor space. Unlike today's computers that operate in binary codes (0's and 1's), the ENIAC operated in decimal (0, 1, 2,..., 9) and it required ten vacuum tubes to represent one decimal digit. With over 18,000 vacuum tubes, the ENIAC needed a great amount of electrical power so much that it was said that it dimmed the lights in Philadelphia whenever it operated.

HUMAN INTELLIGENCE VERSUS MACHINE INTELLIGENCE

Two of the leading mathematicians and computer enthusiasts during the 1900–1950 time frame were Alan Turing and John Von Neumann. In 1945, Von Neumann insisted that computers should not be built as glorified adding machines, with all their operations specified in advance. Rather, he suggested, computers should be built as general-purpose logic machines capable of executing a wide variety of programs. Such machines, Von Neumann proclaimed, would be highly flexible and capable of being readily shifted from one task to another. They could react intelligently to the results of their calculations, could choose among alternatives, and could even play checkers or chess. This represented something unheard of at that time: a machine with built-in intelligence, able to operate on internal instructions.

Prior to Von Neumann's concept, even the most complex mechanical devices had always been controlled from the outside, for example, by setting dials and knobs. Von Neumann did not invent the computer, but what he introduced was equally significant: computing by use of computer programs, the way it is done today. His work paved the way for what would later be called AI in computers.

Alan Turing also made major contributions to the conceptualization of a machine that can be universally used for all problems based only on variable instructions fed into it. Turing's universal machine concept, along with Von Neumann's concept of a storage area containing multiple instructions that can be accessed in any sequence, solidified the ideas needed to develop the programmable computer. Thus, a machine was developed that could perform logical operations and could do them in varying orders by changing the set of instructions that were executed.

Due to the fact that operational machines were now being realized, questions about the "intelligence" of the machines began to surface. Turing's other contribution to the world of Al came in the area of defining what constitutes intelligence. In 1950, he designed the Turing test for determining the intelligence of a system. The test utilized the conversational interaction between three players to try and verify computer intelligence.

The test is conducted by having a person (the interrogator) in a room that contains only a computer terminal. In an adjoining room, hidden from view, a man (Person A) and a woman (Person B) are located with another computer terminal. The interrogator communicates with the couple in the other room by typing questions on the keyboard. The questions appear on the couple's computer screen, and they respond by typing on their own keyboard. The interrogator can direct questions to either Person A or Person B, but without knowing which is the man and which is the woman.

The purpose of the test is to distinguish between the man and the woman merely by analyzing their responses. In the test, only one of the people is obligated to give truthful responses. The other person deliberately attempts to fool and confuse the interrogator by giving responses that may lead to an incorrect guess. The second stage of the test is to substitute a computer for one of the two persons in the other room. Now, the human is obligated to give truthful responses to the interrogator while the computer tries to fool the interrogator into thinking that it is human. Turing's contention is that if the interrogator's success rate in the human/computer version of the game is not better than his success rate in the man/woman version, then the computer can be said to be "thinking." That is, the computer possesses "intelligence." Turing's test has served as a classical example for AI proponents for many years.

By 1952, computer hardware had advanced far enough that actual experiments in writing programs to imitate thought processes could be conducted. The team of Herbert Simon, Allen Newell, and Cliff Shaw organized to conduct such an experiment. They set out to establish what kinds of problems a computer could solve with the right programming. Proving theorems in symbolic logic such as those set forth by Whitehead and Russell in the early 1900s fit the concept of what they felt an intelligent computer should be able to handle.

It quickly became apparent that there was a need for a new, higher-level computer language than was currently available. First, they needed a language that was more user-friendly and could take program instructions that are easily understood by a human programmer and automatically convert them into machine language that could be understood by the computer. Second, they needed a programming language that changed the way in which computer memory was allocated. All previous languages would preassign memory at the start of a program. The team found that the type of programs they were writing would require large amounts of memory and would function unpredictably.

To solve the problem, they developed a list processing language. This type of language would label each area of memory and then maintain a list of all available memory. As memory became available, it would update the list, and when more memory was needed, it would allocate the amount necessary. This type of programming also allowed the programmer to be able to structure his or her data so that any information that was to be used for a particular problem could be easily accessed.

The end result of their effort was a program called Logic Theorist (LT). This program had rules consisting of axioms already proved. When it was given a new logical expression, it would search through all of the possible operations in an effort to discover a proof of the new expression. Instead of using a brute force search method, they pioneered the use of heuristics in the search method.

The LT that they developed in 1955 was capable of solving 38 of 52 theorems that Whitehead and Russell had devised. It was not only capable of the proofs but did them very quickly. What took a LT a matter of minutes to prove would have taken years to do if it had been done by simple brute force on a computer. By comparing the steps that it went through to arrive at a proof to those that human subjects went through, it was also found that it had a remarkable imitation of the human thought process.

Natural Language Dichotomies

Despite the various successful experiments, many observers still believe that AI does not have much potential for practical applications. There is a popular joke in the AI community that points out the deficiency of AI in natural language applications. It is said that a computer was asked to translate the following English statement into Russian and back to English: *The spirit is willing but the flesh is weak.* The reverse translation from Russian to English yielded: *The vodka is good but the meat is rotten.*

From my own author perspective, AI systems are not capable of thinking in the human sense. They are great in mimicking based on the massive amounts of data structures and linkages available. For example, consider the following natural language interpretations of the following ordinary statements:

"No salt is sodium free." A human being can quickly infer the correct interpretation and meaning based on the prevailing context of the conversation. However, an "intelligent" machine may see the same statement in different ways, as enumerated below:

"No (salt) is sodium free," which negates the property of the object, salt. This means that there is no type of salt that is sodium free. In other words, all salts contain sodium.

Alternately, the statement can be seen as follows:

"(No-salt) is sodium free," which is a popular advertisement slogan for the commercial kitchen ingredient named (No-salt). In this case, the interpretation is that this product, named No-salt does not contain sodium.

Here is another one:

"No news is good news." This is a common saying that humans can easily understand regardless of the context. In AI reasoning, it could be subject to the following interpretations:

"(No news) is good news," which agrees with the normal understanding that the state of having no new implies the absence of bad news, which is good (i.e., desirable). In this case, (No-news), as a compound word, is the object.

Or, an AI system could see it as:

"No (news) is good news," which is a contradiction of the normal interpretation. In this case, the AI system could interpret it as a case where all pieces of news are bad (i.e., not good). This implies that the object is the (news).

Here is another one from the political arena:

"The British parliament wants no deal off the table."

The AI interpretations could see the objects as follows:

(No deal), as a condition of negotiation, is off the table.

Alternately, it could be seen, in the negation sense, as all deals are acceptable to be on the table.

Consider the additional examples below:

"Day's days are numbered." To an intelligent machine, the possessive tense, 'Day's' (of a person named Day) could be confused with the plural noun 'Days.'

"The officer found the criminal alone," which could be interpreted in the following two ways:

1. The criminal was alone when found by the officer.
2. The officer was alone when found the criminal.

Pattern recognition is another interesting example that distinguishes human intelligence from machine intelligence. For example, when I park my vehicle in a large shopping center parking lot, with many similarly colored and shaped vehicles, I can always identify my vehicle from a far distance by simply seeing a tiny segment of the body of the vehicle. This could be by seeing half of the headlight jotting out from among several vehicles. It could be by seeing a portion of the tail light. It could even be by seeing the luggage carriage on top of the vehicle barely visible above other vehicles around mine. For an AI system to use pattern recognition to correctly identify my vehicle, it would have to use a tremendous amount of data collection, data manipulation, interpolation, extrapolation, and other complex mathematical algorithms to consider the options to provide a probable match. Considering example such as the above, it can be seen that human intelligence and natural perception still trump machine's simulated intelligence. In spite of this deficiency, machine intelligence, under the banner of AI, can be useful to supplement human intelligence to arrive at a more efficient and effective decision processes. Consequently, AI is a useful and desirable ally in human operations. This belief was what drove the early efforts in defining and advancing the science, technology, engineering, and mathematics foundations for AI.

THE FIRST CONFERENCE ON ARTIFICIAL INTELLIGENCE

The summer of 1956 signified the first attempt to establish the field of machine intelligence into an organized effort. The Dartmouth Summer Conference, organized by John McCarthy, Marvin Minsky, Nathaniel Rochester, and Claude Shannon, brought together people whose work and interest formally founded the field of AI. The conference, held at Dartmouth College in New Hampshire, was funded by a grant from the Rockefeller foundation. It was at that conference that John McCarthy coined the term "artificial intelligence." It was the same John McCarthy who developed the LISP programming language that has become a standard tool for Al development. In attendance at the meeting, in addition to the organizers, were Herbert Simon, Allen Newell, Arthur Samuel, Trenchard More, Oliver Selfridge, and Ray Solomon off.

The LT, developed by Newell, Shaw, and Simon, was discussed at the conference. The system, considering the first Al program, used heuristic search to solve mathematical problems in *Principia Mathematica,* written by Whitehead and Russell (Newell and Simon 1972). Newell and Simon were far ahead of others in actually implementing Al ideas with their LT. The Dartmouth meeting served mostly as an avenue for the exchange of information and, more importantly, as a turning point in the main emphasis of work in the Al endeavor. Instead of concentrating so much on the hardware to imitate intelligence, the meeting set the course for examining the structure of the data being processed by computers, the use of computers to process symbols, the need for new languages, and the role of computers for testing theories.

EVOLUTION OF SMART PROGRAMS

The next major step in software technology came from Newell, Shaw, and Simon in 1959. The program they introduced was called General Problem Solver (GPS). GPS was intended to be a program that could solve many types of problems. It was capable of solving theorems, playing chess, or doing various complex puzzles. GPS was a significant step forward in AI. It incorporates several new ideas to facilitate problem-solving. The nucleus of the system was the use of means-end analysis. Means-end analysis involves comparing a present state with a goal state. The difference between the two states is determined, and a search is done to find a method to reduce this difference. This process is continued until there is no difference between the current state and the goal state.

In order to further improve the search, GPS contained two other features. The first is that, if while trying to reduce the deviation from the goal state, it finds that it has actually complicated the search process, it was capable of backtracking to an earlier state and exploring alternate solution paths. The second is that it was capable of defining subgoal states that, if satisfied, would permit the solution process to continue. In formulating GPS, Newell and Simon had done extensive work studying human subjects and the way they solved problems. They felt that GPS did a good job of imitating the human subjects. They commented on the effort by saying (Newell and Simon, 1961):

The fragmentary evidence we have obtained to date encourages us to think that the General Problem Solver provides a rather good first approximation to an information processing theory of certain kinds of thinking and problem-solving behavior. The processes of "thinking" can no longer be regarded as completely mysterious.

GPS was not without critics. One of the criticisms was that the only way the program obtained any information was to get it from human input. The way and order in which the problems were presented were controlled by humans; thus, the program was doing only what it was told to do. Newell and Simon argued that the fact that the program was not just repeating steps and sequences but was actually applying rules to solve problems it had not previously encountered is indicative of intelligent behavior.

There were other criticisms also. Humans are able to devise new shortcuts and improvise. GPS would always go down the same path to solve the same problem, making the same mistakes as before. It could not learn. Another problem was that GPS was good when given a certain area or a specific search space to solve. The problem with this limitation was that in the solution of problems, it was difficult to determine what search space to use. Sometimes solving the problem is trivial compared with finding the search space. The problems posed to GPS were all of a specific nature. They were all puzzles or logical challenges: problems that could easily be expressed in symbolic form and operated on in a pseudo-mathematical approach. There are many problems that humans face that are not so easily expressed in symbolic form.

Also during the year 1959, John McCarthy came out with a tool that was to greatly improve the ability of researchers to develop AI programs. He developed a new computer programming language called LISP (list processing). It was to become one of the most widely used languages in the field.

LISP is distinctive in two areas: memory organization and control structure. The memory organization is done in a tree fashion with interconnections between memory groups. Thus, it permits a programmer to keep track of complex structural relationships. The other distinction is the way the control of the program is done. Instead of working from the prerequisites to a goal, it starts with the goal and works backward to determine what prerequisites are required to achieve the goal.

In 1960, Frank Rosenblatt did some work in the area of pattern recognition. He introduced a device called PERCEPTRON that was supposed to be capable of recognizing letters and other patterns. It consisted of a grid of 400 photo cells connected with wires to a response unit that would produce a signal only if the light coming off the subject to be recognized crossed a certain threshold.

During the latter part of the 1960s, there were two efforts in another area of simulating human reasoning. Kenneth Colby at Stanford University and Joseph Weizenbaum at MIT wrote separate programs that were capable of interacting in a two-way conversation. Weizenbaum's program was called ELIZA. The programs were able to sustain very realistic conversations by using very clever techniques. For example, ELIZA used a pattern matching method that would scan for keywords such as "I," "you," "like," and so on. If one of these words was found, it would execute rules associated with it. If there was no match found, the program would respond with a request for more information or with some noncommittal response.

It was also during the 1960s that Marvin Minsky and his students at MIT made significant contributions toward the progress of AI. One student, T. G. Evans, wrote a program that could perform visual analogies. The program was shown two figures that had some relationship to each other and was then asked to find another set of figures from a set that matched the same relationship. The input to the computer was not done by a visual sensor (like the one worked on by Rosenblatt), but instead the figures were described to the system.

In 1968, another student of Minsky's, Daniel Bobrow, introduced a linguistic problem solver called STUDENT. It was designed to solve problems that were presented to it in a word problem format. The key to the program was the assumption that every sentence was an equation. It would take certain words and turn them into mathematical operations. For example, it would convert "is" into "=" and "per" into "/."

Even though STUDENTS responded very much the same way that a real student would, there was a major difference in depth of understanding. While the program was capable of calculating the time two trains would collide given the starting points and speeds of both, it had no real understanding or even cared what a "train" or "time" was. Expressions such as "per chance" and "this is it" could mean totally different things than what the program would assume. A human student would be able to discern the intended meaning from the context in which the terms were used.

In an attempt to answer the criticisms about understanding, another student at MIT, Terry Winograd, developed a significant program named SHRDLU. In setting up his program, he utilized what was referred to as a micro-world or blocks-world. This limited the scope of the world that the program had to try to understand. The program communicated in what appeared to be natural language.

The operation of SHRDLU consisted of a set of blocks of varying shapes (cubes, pyramids, etc.), sizes, and colors. These blocks were all set on an imaginary table. Upon request, SHRDLU would rearrange the blocks to any requested configuration. The program was capable of knowing when a request was unclear or impossible. For instance, if it was requested to put a block on top of the pyramid, it would request that the user specify more clearly what block and what pyramid. It could also recognize that the block would not sit on top of the pyramid.

Two other approaches that the program took that were new to programs were the ability to make assumptions and the ability to learn. If asked to pick up a larger block, it would assume a larger block than the one it was currently working on. If asked to build a figure that it did not know, it would ask for an explanation of what it was and, thereafter, it would recognize the object. One major sophistication that SHRDLU added to the science of Al programming was its use of a series of expert modules or specialists. There was one segment of the program that specialized in segmenting sentences into meaningful word groups, a sentence specialist to determine the relationship between nouns and verbs, and a scenario specialist that understood how individual scenes related to one another. This sophistication added much enhancement to the method in which instructions were analyzed.

As sophisticated as SHRDLU was at that time, it did not escape criticism. Other scholars were quick to point out its deficiencies. One of the shortcomings was that SHRDLU only responded to requests; it could not initiate conversations. It also had no sense of conversational flow. It would jump from performing one type of task to a totally different one if so requested. While SHRDLU had an understanding of the tasks it was to perform and the physical world in which it operated, it still could not understand very abstract concepts.

BRANCHES OF ARTIFICIAL INTELLIGENCE

The various attempts to formally define the use of machines to simulate human intelligence led to the development of several branches of Al. Current sub specialties of AI include the following:

1. *Natural language processing:* This deals with various areas of research such as database inquiry systems, story understanders, automatic text indexing, grammar and style analysis of text, automatic text generation, machine translation, speech analysis, and speech synthesis.
2. *Computer vision:* This deals with research efforts involving scene analysis, image understanding, and motion derivation.
3. *Robotics:* This involves the control of effectors on robots to manipulate or grasp objects, locomotion of independent machines, and use of sensory input to guide actions.

4. *Problem-solving and planning:* This involves applications such as refinement of high-level goals into lower-level ones, determination of actions needed to achieve goals, revision of plans based on intermediate results, and focused search of important goals.
5. *Learning:* This area of Al deals with research into various forms of learning including rote learning, learning through advice, learning by example, learning by task performance, and learning by following concepts.
6. *Expert systems:* This deals with the processing of knowledge as opposed to the processing of data. It involves the development of computer software to solve complex decision problems.

NEURAL NETWORKS

Neural networks, sometimes called connectionist systems, represent networks of simple processing elements or nodes capable of processing information in response to external inputs. Neural networks were originally presented as being models of the human nervous system. Just after World War II, scientists found out that the physiology of the brain was similar to the electronic processing mode used by computers. In both cases, large amounts of data are manipulated. In the case of computers, the elementary unit of processing is the bit, which is in either an "on" or "off' state. In the case of the brain, *neurons* perform the basic data processing. Neurons are tiny cells that follow a binary principle of being either in a state of firing (on) or not firing (off). When a neuron is on, it fires a signal to other neurons across a network of synapses.

In the late 1940s, Donald Hebb, a researcher, hypothesized that biological memory results when two neurons are active simultaneously. The synaptic connection of synchronous neurons is reinforced and given preference over connections made by neurons that are not active simultaneously. The level of preference is measured as a weighted value. Pattern recognition, a major strength of human intelligence, is based on the weighted strengths of the reinforced connections between various pairs of simultaneously active neurons.

The idea presented by Hebb was to develop a computer model based on the way in which neurons form connections in the human brain. But the idea was considered to be preposterous at that time since the human brain contains 100 billion neurons and each neuron is connected to 10,000 others by a synapse. Even by today's computing capability, it is still difficult to duplicate the activities of neurons. In 1969, Marvin Minsky and Seymour Pappert wrote the book entitled *Perceptrons,* in which they criticized existing neural network research as being worthless. It has been claimed that the pessimistic views presented by the book discouraged further funding for neural network research for several years. Funding was, instead, diverted to further research of expert systems, which Minsky and Pappert favored. It is only recently that neural networks are beginning to make a strong comeback.

Because neural networks are modeled after the operations of the brain, they hold considerable promise as building blocks for achieving the ultimate aim of AI. The present generation of neural networks use artificial neurons. Each neuron is connected to at least one other neuron in a synapse-like fashion. The networks are based on some form of learning model. Neural networks learn by evaluating changes in input. Learning can be either supervised or unsupervised. In supervised learning, each response is guided by given parameters. The computer is instructed to compare any inputs to ideal responses, and any discrepancy between the new inputs and ideal responses is recorded. The system then uses this data bank to guess how much the newly gathered data is similar to or different from the ideal responses. That is, how closely the pattern matches. Supervised learning networks are now commercially used for control systems and for handwriting and speech recognition.

In unsupervised learning, input is evaluated independently and stored as patterns. The system evaluates a range of patterns and identifies similarities and dissimilarities among them. However, the system cannot derive any meaning from the information without human assignment of values to the patterns. Comparisons are relative to other results, rather than to an ideal result. Unsupervised learning networks are used to discover patterns where a particular outcome is not known in advance, such as in physics research and the analysis of financial data. Several commercial neural network products are now available. An example is NeuroShell from Ward Systems Group. The software is expensive, but it is relatively easy to use. It interfaces well with other software such as Lotus 1–2–3 and dBASE, as well as with C, Pascal, FORTRAN, and BASIC programming languages.

Despite the proven potential of neural networks, they drastically oversimplify the operations of the brain. The existing systems can undertake only elementary pattern-recognition tasks and are weak at deductive reasoning, math calculations, and other computations that are easily handled by conventional computer processing. The difficulty in achieving the promise of neural networks lies in our limited understanding of how the human brain functions. Undoubtedly, to accurately model the brain, we must know more about it. But a complete knowledge of the brain is still many years away.

EMERGENCE OF EXPERT SYSTEMS

In the late 1960s to early 1970s, a special branch of Al began to emerge. The branch, known as expert systems, has grown dramatically in the past few years, and it represents the most successful demonstration of the capabilities of Al. Expert systems are the first truly commercial application of work done in the Al field and as such have received considerable publicity. Due to the potential benefits, there is currently a major concentration in the research and development of expert systems compared with other efforts in Al.

Not driven by the desire to develop general problem-solving techniques that had characterized Al before, expert systems address problems that are focused. When Edward Feigenbaum developed the first successful expert system, DENDRAL, he had a specific type of problem that he wanted to be able to solve. The problem involved determining which organic compound was being analyzed in a mass spectrograph. The program was intended to simulate the work that an expert chemist would do in analyzing the data. This led to the term expert system.

The period of time from 1970 to 1980 saw the introduction of numerous expert systems to handle several functions from diagnosing diseases to analyzing geological exploration information. Of course, expert systems have not escaped the critics. Given the nature of the system, critics argue that it does not fit the true structure of AI. Because of the use of only specific knowledge and the ability to solve only specific problems, some critics are apprehensive about referring to an expert system as being intelligent. Proponents argue that if the system produces the desired results, it is of little concern whether it is intelligent or not.

A controversy of interest surfaced in 1972 with a book published by Hubert Dreyfus called *What Computers Can't Do: A Critique of Artificial Reason.* Views similar to those contained in the book were presented in 1976 by Joseph Weizenbaum. The issues that both authors raised touched on some of the basic questions that prevailed way back in the days of Descartes. One of Weizenbaum's reservations concerned what should ethically and morally be handed over to machines. He maintained that the path that Al was pursuing was headed in a dangerous direction. There are some aspects of human experience, such as love and morality, that could not adequately be imitated by machines.

While the debates were going on over how much Al could do, the work on getting Al to do more continued. In 1972, Roger Shrank introduced the notion of script; the set of familiar events that can be expected from an often encountered setting. This enables a program to quickly assimilate facts. In 1975, Marvin Minsky presented the idea of frames. Even though both concepts did not drastically advance the theory of AI, they did help expedite research in the field.

In 1979, Minsky suggested a method that could lead to a better simulation of intelligence. He presented the "society of minds" view, in which the execution of knowledge is performed by several programs working in conjunction simultaneously. This concept helped to encourage interesting developments such as present-day parallel processing.

As time proceeded through the 1980s, AI gained significant exposure and interest. AI, once a phrase restricted to the domain of esoteric research, has now become a practical tool for solving real problems. While AI is enjoying its most prosperous period, it is still plagued with disagreements and criticisms. The emergence of commercial expert systems on the market has created both enthusiasm and skepticism. There is no doubt that more research and successful applications developments will help prove the potential of expert systems. It should be recalled that new technologies sometimes fail to convince all initial observers. IBM, which later became a giant in the personal computer business, hesitated for several years before getting into the market because the company never thought that those little *boxes* called personal computers would ever have any significant impact on the society. How wrong they were!

The effort in AI is a worthwhile endeavor as long as it increases the understanding that we have of intelligence and as long as it enables us to do things that we previously could not do. Due to the discoveries made in AI research, computers are now capable of things that were once beyond imagination.

Embedded Expert Systems: More expert systems are beginning to show up, not as stand-alone systems, but as software applications in large software systems. This trend is bound to continue as systems integration takes hold in many software applications. Many conventional commercial packages such as statistical analysis systems, data management systems, information management systems, project management systems, and data analysis systems now contain embedded heuristics that constitute expert systems components of the packages. Even some computer operating systems now contain embedded expert systems designed to provide real-time systems monitoring and troubleshooting. With the success of embedded expert systems, the long-awaited payoffs from the technology are now beginning to be realized.

Because the technology behind expert systems has changed little over the past decade, the issue is not whether the technology is useful, but how to implement it. This is why the integrated approach of this book is very useful. The book focuses not only on the technology of expert systems but also on how to implement and manage the technology. Combining neural network technology with expert systems, for example, will become more prevalent. In combination, the neural network might be implemented as a tool for scanning and selecting data while the expert system would evaluate the data and present recommendations.

CONCLUSION

While AI technology is good and amenable to organizational objectives, we must temper it with human intelligence. The best hybrid is when machine intelligence is integrated with human intelligence. The human can handle the intuition part while the machine, as AI, handles the data-intensive and number-crunching parts.

Over the past several years, expert systems have proven their potential for solving important problems in engineering and manufacturing environments. Expert systems are helping major companies to diagnose processes in real time, schedule operations, troubleshoot equipment, maintain machinery, and design service and production facilities. With the implementation of expert systems in industrial environments, companies are finding that real-world problems are best solved by an integrated strategy involving the management of personnel, software, and hardware systems.

Solutions to most engineering and manufacturing problems involve not only heuristics but also mathematical calculations, large data manipulations, statistical analysis, real-time information management, system optimization, and man–machine interfaces. These issues and other related topics are addressed in detail in this book. In addition to the basic concepts of expert systems, guidelines are presented on various items ranging from problem selection, data analysis, knowledge acquisition, and system development to verification, validation, integration, implementation, and maintenance.

Can AI systems and products live up to the hype?

In general, the expectations for all products, systems, and processes of AI include effectiveness, efficiency, ease of use, elegance, safety, security, sustainability, and satisfaction. A systems view can, indeed, bring us closer to realizing these expectations. In terms of a practical and readily seen example of AI, look no further than your mobile phone. The present generation of smart phones is readily a common example of an AI system. So, AI is already all around us on a daily basis.

On a cautionary note, AI is creeping into every facet of our lives. As in managing any technology, AI requires a coordinated management strategy between the software and hardware teams as well as the human end users. AI, via deep fake, has been used to misinform and misdirect the public in recent years, particularly in political contests. So, caution must always be exercised in any default adoption of AI tools and techniques. One product commercial on television proclaims that AI can solve problems previously unsolvable. Using artificial neural networks, AI is being applied, sometimes surreptitiously in the background, to support many functions that we see around us. For example, airlines now use AI to help speed up the boarding process at airports. In addition to the visible physical changes and layouts at airports, there is a preponderance of AI-based technologies operating in the background to facilitate efficiency and effectiveness. One application involves using AI to reduce the time airplanes spend at the gate between flights. Such applications are becoming prevalent in other industries also. In the mix of good and bad, AI will, eventually, become a boon for all. So, we need to get better educated about AI and be ready to leverage it wherever and whenever justified and appropriate.

REFERENCES

Newell, Allen and Herbert A. Simon (1972), *Human Problem-Solving*, Prentice-Hall, Englewood Cliffs, NJ.

Newell, Allen and Herbert A. Simon (1961), "Computer Simulation of Human Thinking," Report, *The RAND Corporation*, Los Angeles, CA, April 20, 1961, p. 2276.

Expert Systems: The Software Side of AI

2

EXPERT SYSTEMS PROCESS

This book is organized in the structure of a strategic process for developing successful expert systems. The strategic process is recommended for anyone venturing into the technology of expert systems from the standpoint of training, research, or applications. This chapter covers the basic concepts of expert systems technology. A basic understanding of these concepts is essential for getting the most out of expert systems. More specific details of the concepts presented in this chapter are discussed in appropriate sections of the subsequent chapters.

EXPERT SYSTEMS CHARACTERISTICS

By definition, an expert system is a computer program that simulates the thought process of a human expert to solve complex decision problems in a specific domain. This chapter addresses the characteristics of expert systems that make them different from conventional programming and traditional decision support tools. The growth of expert systems is expected to continue for several years. With the continuing growth, many new and exciting applications will emerge. An expert system operates as an interactive system that responds to questions,

DOI: 10.1201/9781003089643-2

asks for clarification, makes recommendations, and generally aids the decision-making process. Expert systems provide "expert" advice and guidance in a wide variety of activities, from computer diagnosis to delicate medical surgery.

Various definitions of expert systems have been offered by several authors. A general definition that is representative of the intended functions of expert systems is presented below:

> An *expert system* is an interactive computer-based decision tool that uses both facts and heuristics to solve difficult decision problems based on knowledge acquired from an expert.

An expert system may be viewed as a computer simulation of a human expert. Expert systems are an emerging technology with many areas for potential applications. Past applications range from MYCIN, used in the medical field to diagnose infectious blood diseases, to XCON, used to configure computer systems. These expert systems have proven to be quite successful. Most applications of expert systems will fall into one of the following categories:

- Interpreting and identifying
- Predicting
- Diagnosing
- Designing
- Planning
- Monitoring
- Debugging and testing
- Instructing and training
- Controlling

Applications that are computational or deterministic in nature are not good candidates for expert systems. Traditional decision support systems such as spreadsheets are very mechanistic in the way they solve problems. They operate under mathematical and Boolean operators in their execution and arrive at one and only one static solution for a given set of data. Calculation-intensive applications with very exacting requirements are better handled by traditional decision support tools or conventional programming. The best application candidates for expert systems are those dealing with expert heuristics for solving problems. Conventional computer programs are based on factual knowledge; an indisputable strength of computers. Humans, by contrast, solve problems on the basis of a mixture of factual and heuristic knowledge. Heuristic knowledge, composed of intuition, judgment, and logical inferences, is an indisputable strength of humans. Successful expert systems will be those that combine facts and heuristics and, thus, merge human knowledge with computer power in solving problems. To be effective, an expert system must focus on a particular problem domain as discussed below.

DOMAIN SPECIFICITY Expert systems are typically very domain specific. For example, a diagnostic expert system for troubleshooting computers must perform all the necessary data manipulation as a human expert would. The developer of such a system must limit his or her scope of the system to just what is needed to solve the target problem. Special tools or programming languages are often needed to accomplish the specific objectives of the system.

SPECIAL PROGRAMMING LANGUAGES Expert systems are typically written in special programming languages. The use of languages such as LISP and PROLOG in the development of an expert system "simplifies" the coding process. The major advantage of these languages, as compared with conventional programming languages, is the simplicity of the addition, elimination, or substitution of new rules and memory management capabilities. Presented below are some of the distinguishing characteristics of programming languages needed for expert systems work:

- Efficient mix of integer and real variables
- Good memory management procedures
- Extensive data manipulation routines
- Incremental compilation
- Tagged memory architecture
- Optimization of the systems environment
- Efficient search procedures

EXPERT SYSTEMS STRUCTURE

Complex decisions involve intricate combinations of factual and heuristic knowledge. In order for the computer to be able to retrieve and effectively use heuristic knowledge, the knowledge must be organized in an easily accessible format that distinguishes between data, knowledge, and control structures. For this reason, expert systems are organized into three distinct levels:

1. *Knowledge base:* This consists of problem-solving rules, procedures, and intrinsic data relevant to the problem domain.
2. *Working memory:* This refers to task-specific data for the problem under consideration.
3. *Inference engine:* This is a generic control mechanism that applies the axiomatic knowledge in the knowledge base to the task-specific data to arrive at some solution or conclusion.

These three distinct levels are unique in that the three pieces may very well come from different sources. The inference engine, such as VP-Expert, may come from a commercial vendor. The knowledge base may be a specific diagnostic knowledge base compiled by a consulting firm, and the problem data may be supplied by the end user. A knowledge base is the nucleus of the expert system structure. A knowledge base is not a database. The traditional database environment deals with data that have a static relationship between the elements in the problem domain. A knowledge base is created by knowledge engineers, who translate the knowledge of real human experts into rules and strategies. These rules and strategies can change depending on the prevailing problem scenario. The knowledge base provides the expert system with the capability to recommend directions for user inquiry. The system also instigates further investigation into areas that may be important to a certain line of reasoning, but not apparent to the user.

The modularity of an expert system is an important distinguishing characteristic compared with a conventional computer program. Modularity is affected in an expert system by the use of three distinct components.

The knowledge base constitutes the problem-solving rules, facts, or intuition that a human expert might use in solving problems in a given problem domain. The knowledge base is usually stored in terms of *If–Then rules*. The working memory represents relevant data for the current problem being solved. The inference engine is the control mechanism that organizes the problem data and searches through the knowledge base for applicable rules. With the increasing popularity of expert systems, many commercial inference engines are coming into the market. A survey of selected commercial inference engines is presented in the Appendix at the end of this book. The development of a functional expert system usually centers on the organization of the knowledge base.

A good expert system is expected to grow as it "learns" from user feedback. Feedback is incorporated into the knowledge base as appropriate to make the expert system "smarter." The dynamism of the application environment for expert systems is based on the individual dynamism of the components. This can be classified as follows:

Most Dynamic: **Working Memory**

The contents of the working memory, sometimes called the data structure, change with each problem situation. Consequently, it is the most dynamic component of an expert system, assuming, of course, that it is kept current.

Moderately Dynamic: **Knowledge Base**

The knowledge base need not change unless there is a new piece of information that indicates a change in the problem solution procedure. Changes in the knowledge base should be carefully evaluated before being implemented. In effect, changes should not be based on just one consultation experience. For example, a rule that is found to be irrelevant under one problem situation may turn out to be crucial in solving other problems.

Least Dynamic: **Inference Engine**

Because of the strict control and coding structure of an inference engine, changes are made only if absolutely necessary to correct a bug or to enhance the inferential process. Commercial inference engines, in particular, change only at the discretion of the developer. Since frequent updates can be disrupting and costly to clients, most commercial software developers try to minimize the frequency of updates.

The Need for Expert Systems

Expert systems are necessitated by the limitations associated with conventional human decision-making processes. These limitations include the following:

1. Human expertise is very scarce.
2. Humans get tired from physical or mental workload.
3. Humans forget crucial details of a problem.
4. Humans are inconsistent in their day-to-day decisions.
5. Humans have limited working memory.
6. Humans are unable to quickly comprehend large amounts of data.
7. Humans are unable to retain large amounts of data in memory.
8. Humans are slow in recalling information stored in memory.
9. Humans are subject to deliberate or inadvertent bias in their actions.
10. Humans can deliberately avoid decision responsibilities.
11. Humans *lie, hide,* and die.

Coupled with the abovementioned human limitations are the weaknesses inherent in conventional programming and traditional decision support tools. Despite the mechanistic power of computers, they have certain limitations that impair their effectiveness in implementing human-like decision processes. Some of the limitations of conventional decision support tools are as follows:

1. Conventional programs are algorithmic in nature and depend only on raw machine power.
2. Conventional programs depend on facts that may be difficult to obtain.
3. Conventional programs do not make use of the effective heuristic approaches used by human experts.

4. Conventional programs are not easily adaptable to changing problem environments.
5. Conventional programs seek explicit and factual solutions that may not be possible.

Benefits of Expert Systems

Expert systems offer an environment where the good capabilities of humans and the power of computers can be incorporated to overcome many of the limitations discussed in the previous section. Presented below are some of the most obvious benefits that are offered by expert systems:

1. Expert systems increase the probability, frequency, and consistency of making good decisions.
2. Expert systems help distribute human expertise.
3. Expert systems facilitate real-time, low-cost, expert-level decisions by the nonexpert.
4. Expert systems enhance the utilization of most of the available data.
5. Expert systems permit objectivity by weighing evidence without bias and without regard for the user's personal and emotional reactions.
6. Expert systems permit dynamism through modularity of structure.
7. Expert systems free up the mind and time of the human expert to enable him or her to concentrate on more creative activities.
8. Expert systems encourage investigations into the subtle areas of a problem.

EXPERT SYSTEMS ARE FOR EVERYONE No matter which area of business one is engaged in, expert systems could fulfill the need for higher productivity and reliability of decisions. Everyone can find an application potential in the field of expert systems. Contrary to the belief that expert systems may pose a threat to job security, expert systems can actually help to create opportunities for new job areas. Presented below are some areas that hold promise for new job opportunities. A prospective embracer of expert systems can engage himself or herself in one or more of the following areas:

- Basic research
- Applied research
- Knowledge engineering
- Inference engine development
- Consulting (development and implementation)

- Training
- Sales and marketing
- Passive or active end user

  An active user is one who directly uses expert systems consultations to obtain recommendations. A passive user is one who trusts the results obtained from expert systems and supports the implementation of those results.

Transition from Data Processing to Knowledge Processing

What data has been to the previous generations of computing, knowledge is to the present generation of computing. Expert systems represent a revolutionary transition from traditional data processing to knowledge processing. There is an adaptive relationship between the procedures for data processing and knowledge processing to make decisions. In traditional data processing, the decision-maker obtains the information generated and performs an explicit analysis of the information before making his or her decision. In an expert system, knowledge is processed by using available data as the processing fuel. Conclusions are reached and recommendations are derived implicitly. The expert system offers the recommendation to the decision-maker, who makes the final decision and implements it as appropriate. Conventional data can now be manipulated to work with durable knowledge, which can be processed to generate timely information, which is then used to enhance human decisions.

HEURISTIC REASONING

Human experts use a type of problem-solving technique called heuristic reasoning. This reasoning type, commonly called "rules of thumb" or "expert heuristics," allows the expert to quickly and efficiently arrive at a good solution. Expert systems base their reasoning process on symbolic manipulation and heuristic inference procedures that closely match the human thinking process. Conventional programs can only recognize numeric or alphabetic strings and manipulate them only in a preprogrammed manner.

Search Control Methods: All expert systems are search-intensive. Many techniques have been employed to make these intensive searches more efficient. Branch and bound, pruning, depth first search, and breadth first search are some of the search techniques that have been explored. Because of the intensity of the search process, it is important that good search control strategies be used in expert systems inference processes.

FORWARD CHAINING This method involves checking the condition part of a rule to determine whether it is true or false. If the condition is true, then the action part of the rule is also true. This procedure continues until a solution is found or a dead end is reached. Forward chaining is commonly referred to as data-driven reasoning. Further discussions of forward chaining are presented in subsequent chapters.

BACKWARD CHAINING Backward chaining is used to backtrack from a goal to the paths that lead to the goal. It is the reverse of forward chaining. Backward chaining is very good when all outcomes are known and the number of possible outcomes is not large. In this case, a goal is specified and the expert system tries to determine what conditions are needed to arrive at the specified goal. Backward chaining is also called goal-driven reasoning. More details are provided on the backward chaining process in Chapter 5.

USER INTERFACE

The initial development of an expert system is performed by the expert and the knowledge engineer. Unlike most conventional programs in which only programmers can make program design decisions, the design of large expert systems is implemented through a team effort. A consideration of the needs of the end user is very important in designing the contents and user interface of expert systems.

Natural Language: The programming languages used for expert systems tend to operate in a manner similar to ordinary conversation. We usually state the premise of a problem in the form of a question with actions being stated much as we would verbally answer the question—that is, in a "natural language" format. If during or after a consultation, an expert system determines that a piece of its data or knowledge base is incorrect or is no longer applicable because the problem environment has changed, it should be able to update the knowledge base accordingly. This capability would allow the expert system to converse in a natural language format with either the developers or users.

Expert systems not only arrive at solutions or recommendations but can also give the user a level of confidence about the solution. In this manner, an expert system can handle both quantitative and qualitative factors when analyzing problems. This aspect is very important when we consider how inexact most input data is for day-to-day decision-making. For example, the problems addressed by an expert system can have more than one solution or, in some cases, no definite solution at all. Yet, the expert system can provide useful recommendations to the user just as a human consultant might do.

Explanations Facility in Expert Systems: One of the key characteristics of an expert system is the explanation facility. With this capability, an expert system can explain how it arrives at its conclusions. The user can ask questions dealing with the what, how, and why aspects of a problem. The expert system will then provide the user with a trace of the consultation process; pointing out the key reasoning paths followed during the consultation. Sometimes, an expert system is required to solve other problems, possibly not directly related to the specific problem at hand, but whose solution will have an impact on the total problem-solving process. The explanation facility helps the expert system to clarify and justify why such a digression might be needed.

Data Uncertainties: Expert systems are capable of working with inexact data. An expert system allows the user to assign probabilities, certainty factors, or confidence levels to any or all input data. This feature closely represents how most problems are handled in the real world. An expert system can take all relevant factors into account and make a recommendation based on the "best" possible solution rather than the only exact solution.

Application Roadmap

The symbolic processing capabilities of AI technology lead to many potential applications in engineering and manufacturing. With the increasing sophistication of AI techniques, analysts are now able to use innovative methods to provide viable solutions to complex problems in everyday applications. Figure 2.1 presents a structural representation of the application paths for artificial intelligence (AI) and expert systems. The pathway can vary depending on the specific applications of interest.

SYMBOLIC PROCESSING

Contrary to the practice in conventional programming, expert systems can manipulate objects symbolically to arrive at reasonable conclusions to a problem scenario. The object drawings in this section are used to illustrate the versatility of symbolic processing by using the manipulation of objects to convey information. Let us assume that we are given the collection of five common objects listed below:

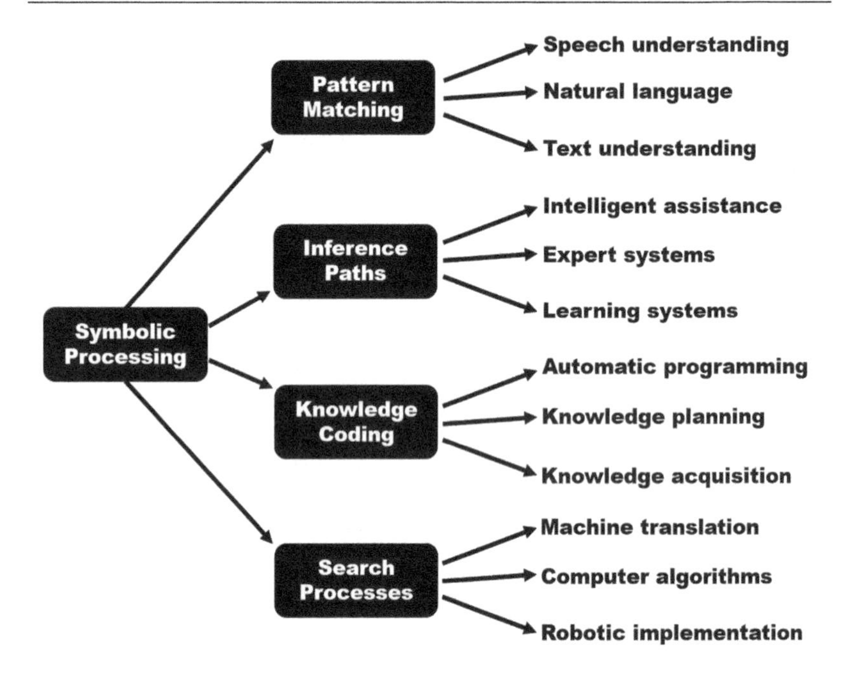

FIGURE 2.1 Application roadmap for expert systems.

Head, Hammer, Bucket, Foot, and Bill (as in doctor's bill).

We can logically arrange a subset of the set of the given objects to convey specific inferences. In one example, four of the five objects are arranged in the order Hammer, Head, Foot, and Bucket. This unique arrangement may be represented by the equation presented below:

Hammer-Head = Foot-Bucket

It is desired to infer a reasonable statement of the information being conveyed by the symbolic arrangement of the objects. An alternate arrangement of another subset (Hammer, Foot, Foot, and Bill) of the given objects is presented below:

Hammer-Foot = Foot-Bill

It is desired to infer a reasonable statement from the above. It should be noted that ordinary mathematical reasoning concerning the equation, Hammer-Foot=Foot-Bill, might lead to Hammer=Bill. However, in AI symbolic reasoning, the context of the arrangement of the objects will determine the proper implication. The reader should attempt to draw the appropriate inferences from the object arrangements before reading the solutions presented below.

If hammer smashes the head, then victim kicks the bucket (i.e., dies). In this case the action part of the statement relates to an action (a fatal one) by the victim of the assault.

If hammer smashes the foot, then assailant foots the bill. In this case the action part of the statement relates to a compensatory action (restitution) by the assailant.

Using a finite set of symbolic objects, we can generate different pieces of information with different permutations of the objects. A particularly interesting aspect of symbolic processing is noted. The object "Foot" conveys one meaning when concatenated with one given object (hammer) and another totally different meaning when concatenated with another object (Bill). In fact, the identification of the object "Bill" is itself symbolically conveyed by the contents of the medical bill. With the capability of symbolic processing, very powerful AI-based tools can be developed for practical applications. However, more research and development efforts will be needed before many of those practical applications can be realized.

FUTURE DIRECTIONS FOR EXPERT SYSTEMS

The intensity of the ongoing efforts in the area of expert systems has created unique opportunities in many spheres of human endeavor. Listed below are some emerging areas:

1. Large-scale research into natural language systems.
2. Further research in knowledge base organization.
3. Development of more efficient search techniques.
4. Insatiable demand for expert-systems-related consulting services.
5. Commercial knowledge bases.
6. More commercial inference engines and development tools.
7. Drive for advanced hardware and software capabilities.
8. Increased need for expert systems training facilities.
9. Continuing growth in the market for expert systems products and services.

The current generation of expert systems represents only the first step in the technology. Not only are there several areas that are yet to be explored, there are also constant improvements to the systems currently in use. With each change in the development of expert systems, the information required from the human expert changes as well. The emerging generation of expert systems are combining shallow or surface knowledge with deep knowledge, with the former being used for routine problem-solving for the purpose of improving efficiency while the latter is used for very difficult problems.

ACADEMIA–INDUSTRY COOPERATION FOR EXPERT SYSTEMS

The interest in AI and expert systems from the standpoints of both research and applications continues to grow widely. A couple of years ago, only a handful of schools offered formal courses in expert systems. But now, several schools have instituted formal expert systems training programs in academic disciplines ranging from industrial engineering, computer science, chemistry, and medicine to psychology. The potential for applying expert systems to a large variety of problems has made the technology very appealing to both practitioners and educators in all areas of applications.

Unique Capability of Academic Institutions: Academic institutions, because of their educational setup, are in a better position to generate, learn, and transfer expert systems knowledge than industry establishments. In fact, most of the significant early developments in expert systems emanated from academic institutions. The insatiable quest for knowledge in academia can fuel the search for innovative solutions to specific industry problems.

Unique Capability of Industry: Industrial establishments are well versed in practical implementation of technology. The commercialization of technology is one impetus that drives further efforts to develop new technology. Technologies that are developed within the academic community mainly for research purposes often languish in laboratories because of the lack of funds or orientation for commercial development. The potentials of these technologies go untapped because of the following:

1. The developer does not know which segment of industry may need the technology.
2. Industry is not aware that the technology to solve their problems is available in some laboratories in some academic institutions.
3. There is no coordinated mechanism for technological interface between industry and academia.

A coordinated marriage between the two bodies (industry and academia) can provide an avenue that facilitates quicker implementation of expert systems technology to solve engineering and manufacturing problems. This will facilitate a smooth relationship between people and technology. Industry has the financial capability, interest, and aggressiveness to bring technologies out of academic laboratories. Also, a cooperative industry is a fertile ground for prototyping new academic ideas.

Industry Need: Many professionals in industry still lack the basic knowledge to successfully take advantage of the capability of expert systems. This is mainly because not many of these professionals had the opportunity to enroll in a formal course in expert systems while they were in school. Many companies are now aggressively urging their employees to attend workshops, conferences, seminars, and formal classes in order to acquire the necessary knowledge. Some companies even organize in-house regular training programs. Despite these efforts, industry professionals still lag behind for three main reasons:

1. Training programs take them away from their regular job functions. Consequently, not enough time is allocated for comprehensive training.
2. Since the payoff on expert systems training may not be immediately apparent, managers tend to want the professional to concentrate his/her effort on a prevailing dollar-valued problem and defer the release time for training.
3. Post-training assignments often do not match the skills acquired from a training. As a result, the professional is unable to implement the new technology skill and finds it difficult to keep abreast of the fast-paced developments in the technology. Retraining is, thus, usually needed when an opportunity for an expert system project finally develops.

Academic Need: Academic institutions interested in expert systems are presently being hampered by the lack of adequate research and training facilities. Industry can help in this regard by providing direct support for selected institutions. The institutions also need real problems to work on as projects or case studies. Industry can provide these under a cooperative arrangement. The development of successful expert systems requires domain experts who are very conversant with the problem environment. Industry professionals can easily fill the role of domain experts for systems developed within the academic environment.

Industry Approach: Because of the limited training opportunity, the industry professional is often caught in a perplexing tangle of requirements that he/she is unable to satisfy. One approach that some companies take in solving this dilemma is to hire fresh graduates who already have the appropriate expert systems training. These younger employees work with experienced professionals who are familiar with companies' operations and problems. This type of arrangement works very well for expert systems projects where the experienced professional serves as the domain expert and the new graduate serves as the knowledge engineer. The problem is that there is still an acute shortage of expert systems graduates to satisfy industry demand.

Academic Approach: Expert systems courses, which are taught in many academic institutions, offer great training opportunities for working professionals who are willing to take courses on a part-time basis. Expert systems courses draw students from many other areas, including engineering, business, mathematics, psychology, meteorology, and music. Many students who graduate with at least one expert systems course listed on their transcripts are aggressively sought by prospective employers. The problem is that these trained students graduate soon after becoming proficient in the new technology. The high turnover rate makes it difficult for schools to maintain a stable team of experienced students to carry on lengthy expert systems projects.

Industry/Academic Interaction: Expert systems courses offered at convenient times during the day facilitate enrollment of industry professionals. The professionals do not necessarily have to miss much time from work in order to participate in course activities. A key requirement of such courses should be a term project that addresses a real-life problem. The industry professionals should be encouraged to select problems that address a prevailing problem within their company. This helps the professionals to have a focused and rewarding effort for the expert systems endeavor. Class projects developed in the academic environment can be successfully implemented in actual work environments to provide tangible benefits. Those benefits include increased productivity, faster response time to company demands, and annual dollar savings.

Continuing Industry Cooperation: Industry-based class projects should not end with the termination of the course. The author always urges the industry professionals to continue to dedicate time to the maintenance of the expert system and to explore other potential applications within the company. This requirement facilitates a continuing interaction between academia and industry even after the course is over. The professionals can consult with the faculty in academia on a continuing basis about new expert-systems-related problems in industry. With this mutually cooperative interaction, new developments in industry are brought to the attention of academia while new academic research developments are discussed with industry professionals. In addition, on-campus students who have no previous industry exposure are given the opportunity to have a "mentor in the real world" through plant tours and informal consultations.

Knowledge Clearinghouse: Academic institutions can serve as convenient locations for knowledge clearinghouses. Such clearinghouses can be implemented under the auspices of expert systems laboratories. Specific expert systems problems in industry can be brought to the laboratories for joint solution not necessarily under a regular class arrangement. This will further enhance the contacts and interactions between professionals in industry and students on campus. The laboratories can serve as repositories for various commercial expert systems tools. These tools will be available to industry professionals to test, learn, and use in the laboratories before deciding to purchase them for in-company implementations. In return, industry could help support the laboratory through the donation of equipment, funds, and personnel time. The services provided by a clearinghouse could include the following:

1. Providing consulting services on expert systems technology to business and industry.
2. Conducting in-plant custom short courses with hands-on projects for business and industry.
3. Serving as a technology library for expert systems information.
4. Providing software, hardware, and technology information services for prospective expert systems developers.
5. Facilitating technology transfer by helping business and industry to move expert systems technology from the laboratory to the marketplace. The transfer can be either of direct expert systems products or of services.
6. Providing technology management guidelines that will enable entrepreneurs to successfully incorporate expert systems products into their existing products and services.
7. Expanding the training opportunities for students and working professionals.

Government Support: The government can support the cooperative interactions between industry and university by providing broad-based funding mechanisms. For example, the National Science Foundation (NSF) recently started to provide funds for Industry/University Cooperative Research Centers. Centers for Artificial Intelligence and Expert Systems hold good potential for funding over the next several years.

SAMPLE OF EXPERT SYSTEMS APPLICATIONS

This section lists a selection of past applications of expert systems developed under the direction and supervision of the author in the 1980s and 1990s. These can provide motivational examples of the diversity of possible applications. These are not References. They are just a listing of examples of past work in the research, development, and applications of expert systems, as documented in journal publications, conference presentations, and graduate thesis research reports. The utility of this list is to demonstrate the fact that expert systems, the decision-support software aspects of AI, have been around for a long time. One recent marketing slogan from a software company touted the fact that their product can solve problems that were "previously unsolvable."

1. Milatovic, M. and A. B. Badiru, "Applied Mathematics Modeling of Intelligent Mapping and Scheduling of Interdependent and Multifunctional Project Resources," ***Applied Mathematics and Computation***, Vol. 149, Issue 3, 2004, pp. 703-721.
2. Milatovic, M. and A. B. Badiru, "Control Sequence Generation in Multistage Fuzzy Control Systems for Design Process," ***AIEDAM (Artificial Intelligence for Engineering Design***, Analysis, and Manufacturing), Vol. 15, 2001, pp. 81-87.
3. Milatovic, Milan and A. B. Badiru, "Fast Estimation of the Modal Position for Unimodally Distributed Data," ***Intelligent Data Analysis***, Vol. 2, No. 1, Jan. 1998,
4. Badiru, A. B. and D. B. Sieger, "Neural Network as a Simulation Metamodel in Economic Analysis of Risky Projects," ***European Journal of Operational Research***, Vol. 105, 1998, pp. 130-142.
5. McCauley-Bell, Pam and A. B. Badiru, "Fuzzy Modeling and Analytic Hierarchy Processing - Means to Quantify Risk Levels Associated with Occupational Injuries -- Part II: The Development of a Fuzzy Rule-Based Model for the Prediction of Injury," ***IEEE Transactions on Fuzzy Systems***, Vol. 4, No. 2, May 1996, pp. 132-138.
6. McCauley-Bell, Pam and A. B. Badiru, "Fuzzy Modeling and Analytic Hierarchy Processing to Quantify Risk Levels Associated with Occupational Injuries -- Part I: The Development of Fuzzy Linguistic Risk Levels," ***IEEE Trans on Fuzzy Systems***, Vol. 4, No. 2, May 1996, pp. 124-131.
7. Badiru, A. B. and Alaa Arif, "Flexpert: Facility Layout Expert System Using Fuzzy Linguistic Relationship Codes," ***IIE transactions***, Vol. 28, No. 4, April 1996, pp. 295-308.
8. Al-Harkan, Ibrahim and A. B. Badiru, "Knowledge-Based Approach to Machine Sequencing," ***Engineering Design and Automation***, Vol. 1, No. 1, Spring 1995, pp. 43-58.
9. Badiru, A. B. and Vassilios Theodoracatos, "Analytical and Integrative Expert System Model for Design Project Management," ***Jnrl of Design and Manufacturing***, Vol. 4, 1994, pp. 195-213.
10. Badiru, A. B., "A New Computational Search Technique for AI Based on Cantor Set," ***Applied Mathematics and Computation***, Vol. 55, 1993, pp. 255-274.

11. Sieger, David B. and A. B. Badiru, "An Artificial Neural Network Case Study: Prediction versus Classification in a Manufacturing Application," ***Computers and Industrial Engineering***, Vol. 25, Nos. 1-4, March 1993, pp. 381-384.
12. Sieger, David B. and A. B. Badiru, "Real-Time Integrated Model for Visual Perception and Fuzzy Control," ***Computers and Industrial Engineering***, Vol. 23, Nos. 1-4, 1992, p. 355-358.
13. Somasundaram, S. and A. B. Badiru, "Intelligent Project Management for Successful Implementation of Continuous Quality Improvement," ***International Journal of Project Management***, Vol. 10, No. 2, May 1992, pp. 89-101.
14. Badiru, A. B., "Successful Initiation of Expert Systems Projects," ***IEEE Transactions on Engineering Management***, Vol. 35, No. 3, August 1988, pp. 186-190.
15. Badiru, A. B., "Cost-Integrated Network Planning Using Expert Systems," ***Project Management Journal***, Vol. 19, No. 2, April 1988, pp. 59-62.
16. Badiru, A. B., Janice Karasz, and Bob Holloway, "AREST: Armed Robbery Eidetic Suspect Typing Expert System," ***Journal of Police Science and Administration***, Vol 16, No 3, Sept 1988, pp. 210-216.
17. Badiru, A. B., "Expert Systems and Industrial Engineers: A Practical Guide for a Successful Partnership," ***Computers & Industrial Engineering***, Vol. 14, No. 1, 1988, pp. 1-13.
18. Milatovic, M.; A. B. Badiru; and T. B. Trafalis, "Taxonomical Analysis of Project Activity Networks Using Competitive Artificial Neural Networks," ***Smart Engineering System Design: Neural Networks. Fuzzy Logic, Evolutionary Programming, Data Mining, and Complex Systems: Proceedings of ANNIE Conference***, ST. Louis, MO, Nov 5-8, 2000.
19. Milatovic, M. and A. B. Badiru, "Intelligent Sectioning for Searching of Unimodal Data," ***Proceedings of 1997 IEEE International Conference on Systems, Man, and Cybernetics***, Orlando, Florida, Oct 12-15, 1997.
20. Milatovic, M. and A. B. Badiru, "Mode Estimating Procedure for Cantor Searching In Artificial Intelligence Systems," ***Proceedings of First International Conference on Engineering Design and Automation***, Bangkok, Thailand, March 1997.
21. Badiru, A. B., V. E. Theodoracatos, and James Grimsley, "State-Space and Expert Systems Hybrid Model for Performance Measurement in Design Integration," ***Proceedings of the 1997 NSF Design & Manufacturing Grantees Conference***, Seattle, WA, January 7-10, 1997, pp. 39-40.

22. Badiru, A. B., V. E. Theodoracatos, and D. B. Sieger, "AI-Based Performance Measurement in Manufacturing Design," ***Proceedings of 5th Industrial Engineering Research Conference***, Minneapolis, MN, May 18-20, 1996, pp. 245-250.
23. Badiru, A. B., V. E. Theodoracatos, and James Grimsley, "State-Space and Expert Systems Hybrid Model for Performance Measurement in Design Integration," ***Proceedings of the 1996 NSF Design & Manufacturing Grantees Conference***, Albuquerque, NM, January 1996.
24. Badiru, A. B., V. Theodoracatos, David Sieger, James Grisley, "State-Space and Expert System Hybrid Model for Performance Measurement in Design Integration," ***Proceedings of the 1995 NSF Design & Manufacturing Grantees Conference***, San Diego, CA, Jan. 4-6, 1995, pp. 85-86.
25. Badiru, A. B. and L. Gruenwald, "A New Approach to AI Database Search Based on Cantor Set," ***Proceedings of International Conference on Computer Applications in Industry and Engineering***, Honolulu, Hawaii, December 15- 17, 1993, pp. 195-199.
26. Chandler, D., G. Abdelnour, S. Rogers, J. Huang, A. Badiru, J. Cheung, C. Bacon, "Fuzzy Rule-Based AI System for Early Fault Detection Prediction," ***Proceedings of the 7th Oklahoma Symposium on Artificial Intelligence,*** Stillwater, Oklahoma, November 18-19, 1993, pp. 62-66.
27. Wei, Hong; L. Gruenwald; and A. B. Badiru, "Improving AI Cantor Set Search for Applications in Database and Artificial Intelligence," Proceedings of the 7th Oklahoma Symposium on Artificial Intelligence, Stillwater, Oklahoma, November 18-19, 1993, pp. 250-259.
28. Oliver, Gaugarin E. and A. B. Badiru, "An Expert System Model for Supplier Development Program in a Manufacturing Firm," Proceedings of the 7th Oklahoma Symposium on Artificial Intelligence, Stillwater, Oklahoma, November 18-19, 1993, pp. 135-141.
29. Badiru, A. B., "Search Heuristic for Expert Systems in Manufacturing Design," Proceedings of First Africa-USA International Conference on Manufacturing Technology, Lagos, Nigeria, January 11-14, 1993, pp. 259-266.
30. Rogers, Steven Hill and A. B. Badiru, "A Fuzzy Set Theoretic Framework for Knowledge-Based Simulation," Proceedings of 15th Annual Conference on Computers and Industrial Engineering, Cocoa Beach, Florida, Vol. 25, Nos. 1-4, March 1993, pp. 119-122.
31. McCauley-Bell, Pamela and A. B. Badiru, "Concept Mapping as a Knowledge Acquisition Tool in the Development of a Fuzzy

Rule-Based Expert System," Proceedings of 15th Annual Conference on Computers and Industrial Engineering, Cocoa Beach, Florida, Vol. 25, Nos. 1-4, March 1993, pp. 115-118.
32. Sieger, Dave and A. B. Badiru, "Neural Network as a Simulation Metamodel in Economic Analysis of Risky Projects," presented at the ORSA/TIMS conference, Phoenix, Arizona, October 31 - November 3, 1993.
33. Badiru, A. B.; Le Gruenwald; and Theodore Trafalis, "A New Search Technique for Artificial Intelligence Systems," Proceedings of the Sixth Oklahoma Symposium on Artificial Intelligence, Tulsa, Oklahoma, November 11-12, 1992, pp. 91-96.
34. Arif, Alaa E. and A. B. Badiru, "An Integrated Expert System with a Fuzzy Linguistic Model for Facilities Layout," Proceedings of the Sixth Oklahoma Symposium on Artificial Intelligence, Tulsa, Oklahoma, November 11-12, 1992, pp. 185-194.
35. Rogers, Steve and A. B. Badiru, "AI Fuzzy Reliability Modeling," Proceedings of IASTED International Conference on Reliability, Quality Control, and Risk Assessment, Washington, DC, November 4-6, 1992, pp. 38-40.
36. McCauley-Bell, Pamela and A. B. Badiru, "An AI Fuzzy Linguistics Model for Job Related Injury Risk Assessment," Computers & Industrial Engineering (Proceedings of 14th Annual Conference on Computers and Industrial Engineering, Orlando, Florida, March 1992), Vol. 23, Nos. 1-4, 1992, pp. 209-212.
37. Badiru, A. B. and Shivakumar Raman, "An Integrative Approach to Designing Expert Systems for Robots Selection," Proceedings of SME Fourth World Conference on Robotics Research, Pittsburgh, PA, September 17-19, 1991, pp. 13.15-13.28.
38. Somasundaram, S. and A. B. Badiru, "An Expert System for External Cylindrical Grinding: Planning, Diagnosing, and Training," Proceedings of the Fifth Oklahoma Symposium on Artificial Intelligence, Norman, Oklahoma, November 1991, pp. 20-29.
39. Badiru, A. B., "OKIE-ROOKIE: An Expert System for Industry Relocation Assessment in Oklahoma, "Proceedings of the Fifth Oklahoma Symposium on Artificial Intelligence, Norman, Oklahoma, November 1991.
40. Chetupuzha, Joseph M. and A. B. Badiru, "AI Design Considerations for Knowledge Acquisition," Proceedings of 13th Annual Conference on Computers and Industrial Engineering, Orlando, Florida, March 1991, Computers & Industrial Engineering, Vol. 21, Nos. 1-4, 1991, pp. 257-261.

41. Karasz, Janice; Bob Holloway; and A. Badiru, "AI-Based Writing Skills for Technical Academia Using Computers," Proceedings of 13th Annual Conference on Computers and Industrial Engineering, Orlando, Florida, March 1991, Computers & Industrial Engineering, Vol. 21, Nos. 1-4, 1991, pp. 407-411.
42. Badiru, A. B., "Justification of Expert Systems Using Analytic Hierarchy Process," Proceedings of 1991 World Congress on Expert Systems, Orlando, Florida, December 1991.
43. Sunku, Ravindra and A. B. Badiru, "ROBEX (Robot Expert): An Expert System for Manufacturing Robot System Implementation," Proceedings of 12th Annual Conference on Computers and Industrial Engineering, Orlando, Florida, March 1990, Computers & Industrial Engineering, Vol. 19, Nos. 1-4, 1990, pp. 481-483.
44. Badiru, A. B., "AI State Space Modeling for Knowledge Representation in Project Monitoring and Control," presented at the ORSA/TIMS Fall Conference, Denver, October 1988.
45. Badiru, A. B. and Hassan Haideri, "Use of Expert Systems in the Heat Treatment of Steel," presented at the ORSA/TIMS Spring Conference, Washington, DC, April 1988.
46. Holloway, Bob; Janice Karazs; and A. B. Badiru, "Knowledge Elicitation for Expert Systems in the Law Enforcement Domain," Proceedings of 11th Annual Conference on Computers and Industrial Engineering, Orlando, Florida, March 1989, Computers & Industrial Engineering, Vol. 17, Nos. 1-4, 1989, pp. 90-94.
47. Joshi, Ajay P.; Neetin N. Datar; and A. B. Badiru, "Knowledge Acquisition and Transfer," in Proceedings of the 1988 Oklahoma Symposium on Artificial Intelligence, Norman, Oklahoma, November 1988, pp. 355-378.
48. Datar, Neetin N. and A. B. Badiru, "A Prototype Knowledge Based Expert System for Robot Consultancy – ROBCON," in Proceedings of the 1988 Oklahoma Symposium on Artificial Intelligence, Norman, Oklahoma, November 1988, pp. 51-68.
49. Badiru, A. B., Janice M. Mathis, and Bob T. Holloway, "Knowledge Base Design for Law Enforcement," Proc of 10th Annual Conf on Computers and Industrial Engineering, Dallas, Texas, March 1988, Computers & Industrial Engineering, Vol. 15, Nos. 1-4, 1988, pp. 78-84.
50. Badiru, A. B., "Expert Systems Application in Manufacturing," seminar, Center of Excellence in Manufacturing, Tennessee Technological University, August 14, 1988.

51. Badiru, A. B., "Applications of Artificial Intelligence in Economic Development," Oklahoma State Legislature, February 1987.
52. Badiru, A. B., "AI Cantor Set Modeling for Manufacturing Knowledge Representation," presented at the ORSA/TIMS Spring Conference, New Orleans, April 1987.
53. Badiru, A. B. and Gary E. Whitehouse, "The Impact of the Computer on Resource Allocation Algorithms," Presented at the ORSA/TIMS Fall Conference, Miami, Florida, October 1986.
54. Badiru, A. B., "Expert Systems," Invited Lecture, Health Sciences Center, Oklahoma City, March 1986.
55. Badiru, A. B. and James R. Smith, "Setting Tolerances by Intelligent Computer Simulation," Proceedings of 1982 IIE Fall Conference, Cincinnati, Ohio, November 1982, pp. 284-288.
56. Bridges, Tim, "The Effect of Intermittent Forgetting Upon Learning and Productivity Within Production Systems," Ph.D. Dissertation, University of Oklahoma, 2000. ***Professor of Operations Management, University of Central Oklahoma.***
57. Milatovic, Milan, "Mapping of Multicapable and Interdependent Resource Units in PERT/CPM Networks" Ph.D. Dissertation, University of Oklahoma, 2000. ***Won the IIE E. J. Sierleja Fellowship Award in 1999. Research & Development Scientist, Mercaritech Corp. Arkansas. Voted one of Professional Top 40 Under 40 in Northwest Arkansas, 2001.***
58. Sieger, David, "Performance Quantification Model Using State Space for Design and Manufacturing Integration," Ph.D. Dissertation, University of Oklahoma, 1995. ***Lead Software Engineer, Hyperfeed Technologies, Chicago. Formerly assistant professor of IE, University of Illinois at Chicago.***
59. Rogers, Steve, "Fuzzy Inferencing Approach to Modeling and Simulation," Ph.D. Dissertation, University of Oklahoma, 1993. ***Senior Artificial Intelligence Engineering, Seagate Technologies, Oklahoma City.***
60. McCauley-Bell, Pamela, "A Fuzzy Linguistic Artificial Intelligence Model for Assessing Risks of Cumulative Trauma Disorders of the Forearm and Hand," Ph.D. Dissertation, University of Oklahoma, 1993.
61. Peters, John, "Generating Intelligent Knowledge Base Information from MIS Data Bases for Equipment Diagnostic Expert Systems," Ph.D. Dissertation, University of Oklahoma, 1992. ***Mining Operations Manager (retired), Morrison Knudsen, Inc., Las Vegas.***

62. Nsofor, Godswill, "A Comparative Analysis of AI Predictive Data-Mining Techniques," MS Thesis, Department of Industrial & Information Engineering, University of Tennessee, 2006.
63. Downes, Paula Sue, "Development of An Intelligent Transportation Network Model for Complex Economic and Infrastructure Simulations," M.S. Thesis, Department of Industrial & Information Engineering, University of Tennessee, 2006
64. Delgado, Vincent G., "An Intelligent Computer Modeling Approach for Critical Resource Diagramming Network Analysis in Project Scheduling," M.S. Thesis, Department of Industrial & Information Engineering, University of Tennessee, 2004.
65. Gunter, Jamie Ehresman, "Intelligent Implementation of Critical Path Method and Critical Resource Diagramming Using Arena Simulation Software," M.S. Thesis, Department of Industrial & Information Engineering, University of Tennessee, 2004.
66. Milatovic, Milan, "Development of Mode Estimating Technique for Cantor Search of Sorted Data in Artificial Intelligence Systems and Manufacturing Design," M.S. Thesis, School of Industrial Engineering, University of Oklahoma, 1996.
67. Fox, Linda, "The Application of Expert System Qualitative Information to Long Range Forecasts," M.S., School of Industrial Engineering, University of Oklahoma, 1995.
68. Benzo, Catherine F., "Expert System Simulation Metamodel for Shop Floor Performance Analysis," M.S., School of Industrial Engineering, University of Oklahoma, 1993.
69. Oliver, Gaugarin, "An Expert System Model for Supplier Development Program," M.S., School of Industrial Engineering, University of Oklahoma, 1993.
70. Baxi, Herschel J., "A Prototype Expert System for Integrated Project Management," M.S., School of Industrial Engineering, University of Oklahoma, 1993.
71. Nakada, Masayuki, "AI Experimental Investigation of Alternate Search Key Distributions for Cantor Set Search Algorithm," M.S., School of Industrial Engineering, University of Oklahoma, 1993.
72. Wei, Hong, "Improving AI Cantor Set Search for Applications in Database and Artificial Intelligence," M.S. (Computer Science), University of Oklahoma, 1993 (co-chair with Dr. Le Gruenwald).
73. Arif, Alaa E., "An Integrated Expert System with a Fuzzy Linguistic Model for Facilities Layout," M.S., School of Industrial Engineering, University of Oklahoma, 1993.

74. Maganty, Radha, "Fuzzy Expert System Model for Assessment of Quality Level in a Company Using MBNQA Criteria," M.S., School of Industrial Engineering, University of Oklahoma, 1993.
75. Sieger, David Bruce, "A Methodology for A Real-Time Artificial Intelligence Surveillance System," M.S., School of Industrial Engineering, University of Oklahoma, 1993.
76. Muppavarapu, Krishna K., "Quality Inspection of Automotive Engine Valves Using AI Knowledge-Based Computer Vision," M.S., School of Industrial Engineering, University of Oklahoma, 1992.
77. Simha, Arun, "An Expert System Model for Pareto Analysis," M.S., School of Industrial Engineering, University of Oklahoma, 1992.
78. Chetupuzha, Joseph, "AI Design Considerations for Knowledge Acquisition and Multiple Knowledge Integration in Expert Systems," M.S., School of Industrial Engineering, University of Oklahoma, 1992.
79. B, Rajesh Nath, "PC Opal: AI-based PC Tool for Optimizing Parameter Levels: An Expert System to Design Experiments Using Taguchi's Orthogonal Arrays and to Analyze Responses," M.S., School of Industrial Engineering, University of Oklahoma, 1991.
80. Sunku, Ravindra, "ROBEX (Robot Expert): An Expert System for Manufacturing Robot System Implementation," M.S., School of Industrial Engineering, University of Oklahoma, 1991.
81. Karode, Amol W., "An Integrated Approach for Multiple AI Knowledge Representation: Hierarchical Blackboard- Based Expert Statistical Process Control System," M.S., School of Industrial Engineering, University of Oklahoma, 1991.
82. Sundaram, Deepak "JUSTEX: An Expert System for the Justification of Advanced Manufacturing Technology," M.S., School of Industrial Engineering, University of Oklahoma, 1991.
83. Nowland, Russell, "Design of Multimedia Expert System for Design for Manufacturability," M.S., School of Industrial Engineering, University of Oklahoma, 1989.
84. Dhanuskodi, Satyanarayanan, "An Expert System for Simulation Modeling of Project Networks," M.S., School of Industrial Engineering, University of Oklahoma, 1989.
85. Joshi, Ajay, "PROCESS-PLUS: A Prototype Expert System for Generative Process Planning," M.S., School of Industrial Engineering, University of Oklahoma, 1989.
86. Datar, Neetin, "A Prototype Expert System for Robot Consultancy (ROBCON)," M.S., School of Industrial Engineering, University of Oklahoma, 1988.

87. Shore, Stan, "AI Systematic Approach to Design Constraint Convergence when Integrating Manufacturing Facilities," M.S., School of Industrial Engineering, University of Oklahoma, 1988.
88. Khuzema, K., "Expert System for Heuristic Selection for Project Scheduling," M.S., School of Industrial Engineering, University of Oklahoma, 1988.
89. Kelley, Scott Douglas, "Design of a framework for an expert system to manage electromagnetic interference in the hospital environment," M.S. Thesis, University of Oklahoma, 1997.
90. Arviso, Wynette R., "Investigation of the spoken Navajo language in the development of an effective human-computer interface," M.S. Thesis, University of Oklahoma, 1996.
91. Chowdhury, A. K. M. Moniruzzaman, "Expert System for Prioritization of Contaminant Sources in Wellhead Protection Areas," Ph.D. Dissertation (Civil Engineering & Environmental Science), University of Oklahoma, 1995.
92. Mathis, Janice, "Development of AI-Based Technical Knowledge for Writing Across the Curriculum for Education Majors," Ph.D. Dissertation (College of Education), University of Oklahoma, 1994.
93. Johnson, Oren, "Artificial Intelligence Decision Support System for Emergency Notification for Tornado Incidents," Ph.D. Dissertation (College of Business), University of Oklahoma, 1993.
94. Chen, Jacob Jen-gwo, "Prototype expert system for physical work stress analysis," Ph.D. Dissertation, University of Oklahoma, 1987.
95. Kesavan, Srikumar, "Dynamic simulation of flexible manufacturing expert system using a process controller and modular component," M.S. Thesis, University of Oklahoma, 1987.

Digital Systems Framework for AI

3

DIGITAL FRAMEWORK FOR AI

In this chapter, we discuss how AI has advanced rapidly in the digital era. Many of the discussions are in the context of the concepts, tools, and techniques of systems engineering (SE), which is a discipline dedicated to integrating elements to arrive at a more robust whole. From an artificial intelligence (AI) implementation perspective, the digital era consists of digital-based science, technology, engineering, and mathematics (STEM). A digital framework is essential for implementing AI because of the desire to achieve integrated capabilities faster, more efficiently, more effectively, more adaptively, more consistently, and more resource-consciously. Some of the considerations needed for such and integrated environment include the following:

- IT Infrastructure for computing needs
- Centralized cloud system for secure data storage
- Collaborative ecosystem of models, tools, techniques, and strategies
- Availability and accessibility of reliable data
- Open architectures to facilitate compliance with standards
- Integrated processes that are sustainable throughout the system life cycle
- Agility of the workforce to embrace and utilize digital platforms

DOI: 10.1201/9781003089643-3

DIGITAL ENGINEERING AND SYSTEMS ENGINEERING

Systems function better when approached from a digital platform. This nexus paves the way for better AI implementations. This requires the humans in the loop of the process to also have a digital mindset. A digital tool that is devoid of the digital readiness of humans will not be sustainable. So, workforce development along the digital spectrum is essential for the embrace of AI. Overlaying the above understanding of a digital environment on the common definition of industrial and systems engineering (ISE), we see how a digital systems framework fits the expectations, goals, and objectives of the digital era (Badiru, 2014a; Badiru, 2014b). For topical relevance in this section, the following definitions apply:

Definition of a system: A system is a collection of interrelated elements, whose collective output (together) is higher than the sum of the individual elements of the system.

Definition of industrial engineering: Industrial engineering is the profession concerned with the design, installation, and improvement of integrated systems of people, materials, information, equipment, and energy by drawing upon specialized knowledge and skills in the mathematical, physical, and social sciences, together with the principles and methods of engineering analysis and design to specify, predict, and evaluate the results to be obtained from such systems.

Definition of systems engineering: SE is the application of engineering tools and techniques to the solutions of multifaceted problems through a systematic collection and integration of parts of the problem with respect to the life cycle of the problem. It is the branch of engineering concerned with the development, implementation, and use of large or complex systems. It focuses on specific goals of a system considering the specifications, prevailing constraints, expected services, possible behaviors, and structure of the system. It also involves a consideration of the activities required to ensure that the system's performance matches specified goals.

Definition of digital engineering: Digital engineering is the combined art and science of creating, capturing, designing, evaluating, justifying, and integrating data using digital (i.e., electronic) tools and processes.

Definition of digital systems framework: Digital systems framework, as presented in this book, is the process of using the hybrid processes of industrial engineering, SE, and digital engineering to manage, allocate, and organize resources to achieve organizational and operational goals in an efficient, effective, sustainable, and repeatable manner. The digital aspect facilitates repeatability and consistency.

SE addresses the integration of tools, people, and processes required to achieve a cost-effective and timely operation of the system. Of importance is the systematic linking of inputs to goals and outputs, explicit treatment of the integration of tools, people, and operational processes. A typical decision support model is a representation of a system, which can be used to answer questions about the system. While SE models facilitate decisions, they are not typically the conventional decision support systems. The end result of using a SE approach is to integrate a solution into the normal organizational process. For that reason, the DEJI systems model® is desired for its structured framework of design, evaluation, justification, and integration. This is essential for digital systems implementation and AI applications.

INTRODUCTION TO DEJI SYSTEMS MODEL

Systems performance is at the intersection of efficiency, effectiveness, and productivity. Efficiency provides the framework for performance in terms of resources and inputs required to achieve the desired level of quality. Effectiveness comes into play with respect to the application of performance to meet specific needs and requirements of an organization. Productivity is an essential factor in the pursuit of digital implementation of AI system as it relates to the throughput of an organization. To achieve the desired levels of quality, efficiency, effectiveness, and productivity, a new research framework must be adopted. In this section, we present a performance enhancement model for system design, evaluation, justification, and integration (DEJI). The model is relevant for research efforts in digital SE and AI applications. The DEJI model of SE provides one additional option for a digital systems platform. Although the model is generally applicable in all types of systems modeling, systems quality is specifically used to describe how the DEJI model is applied. The core stages of the DEJI model are as follows:

- Design
- Evaluation
- Justification
- Integration

Design encompasses any system initiative providing a starting point for a project. Thus, design can include technical product design, process initiation, and concept development. In essence, we can say that "design" represents requirements and specifications. Evaluation can use a variety of metrics both qualitative and quantitative, depending on the organization's needs. Justification can be done on the basis of monetary, technical, or social reasons. Integration needs to be done with respect to the normal or standard operations of the organization. Figure 3.1 illustrates the full profile of the DEJI model.

All the operational elements embedded in the DEJI model are explained and described as presented below:

Design embodies agility, defines end goal, and engages stakeholder.
Evaluate embodies feasibility, metrics, gathers evidence, and assesses utility.
Justify embodies desirability, focus on implementation, and articulate conclusions.

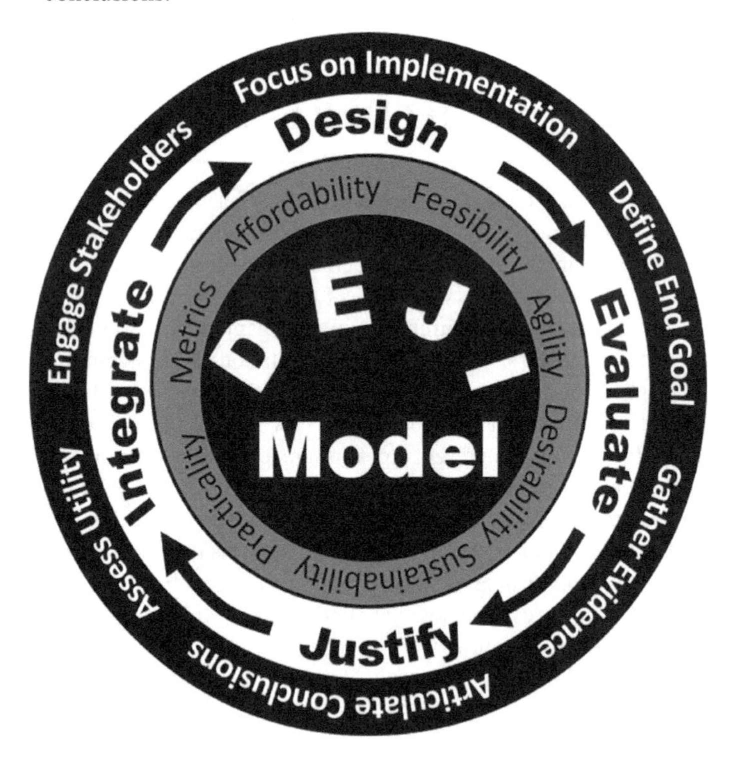

FIGURE 3.1 DEJI systems model for design, evaluation, justification, and integration.

Integrate embodies affordability, sustainability, and practicality.
For application purposes, these elements interface and interact systematically to enhance overall operational performance of an organization.

Application of DEJI Systems Model to Systems Quality

Several aspects of quality must undergo rigorous research along the realms of both quantitative and qualitative characteristics. Many times, quality is taken for granted and the flaws only come out during the implementation stage, which may be too late to rectify. The growing trend in product recalls is a symptom of a priori analysis of the sources and implications of quality at the product conception stage. This approach advocates the use of the DEJI model for enhancing quality design, quality evaluation, quality justification, and quality integration through hierarchical and stage-by-stage processes. Better digital quality is achievable, and there is always room for improvement in the quality of products and services. But we must commit more efforts to the research at the outset of the system development cycle. Even the human elements of the perception of quality can benefit from more directed research from social and behavioral sciences point of view, in a way that will benefit the acceptance of AI applications. Specifically, the DEJI systems model, because of its focus on integration, can address the following systems objectives effectively:

- Creation of an integrated environment provides overarching guidance to influence corporate culture and digital implementation
- Provision of digital tools for empowerment and collaboration to facilitate analysis, data visualization, and policy implementation
- Provision of guidance for the implementation policies and procedures
- Development of consistent IT support system for digital transformation
- Institution of workforce development programs and training to keep pace with the digital environment

There are several other SE models suitable for building a digital systems framework for AI. Some of the well-known models (Badiru, 2019) are discussed next.

The Waterfall Model

The waterfall model, also known as the linear-sequential life-cycle model, breaks down the SE development process into linear sequential phases that do not overlap one another. The model can be viewed as a flow-down approach to engineering development. The waterfall model assumes that each preceding phase must be completed before the next phase can be initiated. Additionally, each phase is reviewed at the end of its cycle to determine whether or not the project aligns with the project specifications, needs, and requirements. Although the orderly progression of tasks simplifies the development process, the waterfall model is unable to handle incomplete tasks or changes made later in the life cycle without incurring high costs. This makes sense for the waterfall model since water normally flows downward, unless forced to go upward through a pumping device, which could be an additional cost. Therefore, this model lends itself better to simple projects that are well defined and understood.

Case Study: This model was instrumental to updating the Conway Regional Medical Center in Arkansas. In 2010, the regional center still did not have an electronic database and information management system for its home healthcare patients. Consequently, the hospital applied the waterfall method to their acquisition of software to handle their needs. First, the hospital management defined their problem as needing a way to maintain a database of documentation and records for the home healthcare patients. The hospital then elected to buy, rather than design, the software necessary for this project. After defining their system requirements, the hospital's administration team purchased what they evaluated to be the most suitable software option. However, the hospital performed systems testing before integrating the software into the home healthcare system. Upon completion of testing, the hospital found that the code needed to be updated once every 6 weeks. This update was factored into their operation and maintenance plan for use of this new system. The system was finally deemed a resounding success once the system was implemented, tested, and the operations and maintenance schedules were created. Through the use of the waterfall model, the software got on track for installation as the primary home healthcare software for the Conway Regional Medical Center.

The V-Model

The V-model, or the verification and validation model, is an enhanced version of the waterfall model that illustrates the various stages of the system life cycle. It is, perhaps, the most used SE model. The V-model is similar to the waterfall model in that they are both linear models whereby each phase is verified before moving on to the next phase. Beginning from the left side, the V-model depicts the development actions that flow from concept of operations to the integration and verification activities on the right side of the diagram. With this model, each phase of the life cycle has a corresponding test plan that helps identify errors early in the life cycle, minimize future issues, and verify adherence to project specifications. Thus, the V-model lends itself well to proactive defect testing and tracking. However, a drawback of the V-model is that it is rigid and offers little flexibility to adjust the scope of a project. Not only is it difficult, but it is also expensive to reiterate phases within the model. Therefore, the V-model works best for smaller tasks where the project length, scope, and specifications are well defined.

Case Study: The V-model was used with great success to bring the Chattanooga Area Regional Transportation Authority (CARTA), which was one of the first smart transport systems in the United States. The V-model was used to guide the design of the new system and to integrate this into the existing system of buses, electric transport, and light rail cars. The new smart system introduced a litany of features such as customer data management, automated route scheduling to meet demand, automated ticket vending, automated diagnostic maintenance system, and computer-aided dispatch and tracking. These features were revolutionary for a mid-sized metropolitan area. CARTA was able to maintain their legacy transport, while integrating their new system. They were able to do this by splitting the V-model into separate sections. CARTA had a dedicated team to manage the flat portions of the V. This portion encapsulated the legacy transportation systems. CARTA then had individual teams focus on the definition, test, and integration of all the new components of the system. Separating the two sides—legacy and innovation—enabled CARTA to maintain valuable functionalities while adding new features to their system that enhanced usability, safety, and efficiency.

Spiral Model

The Spiral model is similar to the V-model in that it references many of the same phases as part of its color-coordinated slices, which indicates the project's stage of development. This model enables multiple flows through the cycle to build a better understanding of the design requirements and engineering complications. The reiterative design lends itself well to supporting popular model-based SE techniques such as rapid prototyping and quick failure methods. Additionally, the model assumes that each iteration of the spiral will produce new information that will encourage technology maturation, evaluate the project's financial situation, and justify continuity. Consequently, the lessons learned from this method provide a data point with which to improve the product. Generally, the spiral model meshes well with the defense life-cycle management vision and integrates all facets of design, production, and integration.

The spiral model is the foundation of the RQ-4 Global Hawk Operational Management and Usage platform. The Global Hawk was phased into operation in six distinct spirals, with each spiral adding new capabilities to the airframe. The first spiral was getting the aircraft in the sky and having a support network to keep it in the air. Everything from pilots to maintainers was optimized to keep the Global Hawk in the air as much as possible. The subsequent phases added imagery (IMINT), signal (SIGINT), radar, and survivability capabilities to the airframe. Each of these capabilities was added one at a time in a spiral development cycle to ensure that each one was integrated into the airframe to the operational standard and could be adjusted to meet this standard before moving on to the next capability. The benefit of incrementally adding capabilities in a spiral fashion greatly helped the Global Hawk stay on budget and schedule for operational rollout.

Defense Acquisitions University SE Model

The new Defense Acquisitions University (DAU) model for SE also originates from the V-model. However, unlike the traditional V-model, it allows for process iteration similarly to the spiral model. A unique attribute of the DAU process is that its life cycle does not need to be completed in order to gain the benefit of iteration. Whereas the spiral model requires the life-cycle process to be completed, the DAU model can refine and improve products at any point in its phase progression. This design is beneficial to making early-stage improvements, which helps systems engineers to avoid budgeting issues such as cost overruns. Moreover, the model allows for fluid transition between project definition (decomposition) and product completion (realization), which is useful in software production and integration. Overall, the DAU model is a fluid combination of the V-model and spiral model.

This tailored V-model was used by the Air Force researchers to create a system to aid Battlefield Airmen in identifying friendly forces and calling in close air support with minimal risk to ground troops. The model was used to find an operational need and break it down into a hierarchy of objectives. The researchers used the hierarchy to design multiple prototypes that attempted to incorporate all of the stated objectives. They used rapid prototyping methods to produce these designs, and they were then tested operationally within Battlefield Airmen squadrons. Ultimately, the production of a friendly marking device was achieved, and this valuable capability was able to be delivered to the warfighter.

Walking Skeleton Model

The walking skeleton model is a lean approach to incremental development, popularly used in software design. It centers on creating a skeleton framework for what the system is going to do and look like. This basic starting point of the system will have minimal functionality, and the systems engineer will work to add muscles to the skeleton. The first step creates a system that may do a very basic yet integral part of the final system design. For example, if one were to design a car using this method, the skeleton would be an engine attached to a chassis with wheels. Once the first basic step is done, the muscles begin to be added to the skeleton. These muscles are more refined and are added one at a time, meaning that each new feature of the system must be completed to add. Furthermore, it is highly recommended that the most difficult features of the system are the first muscles to be added. System components that take a lot of time, require contracting/outsourcing, or are the primary payload must be the first to be completed. This will become the heart of the skeleton, and the rest of the architecture can be optimized to ensure that the critical capability of the system is preserved and enhanced.

An example of engineering using the walking skeleton model is the Boston Dynamics walking dogs. The first thing that the engineers did for these robotic creatures was to create a power source and mobility framework. From there, the engineers were able to then go piece by piece and add more functionality to the project, such as the ability to open doors, pick up objects, and even carry heavy loads. The lessons they learned from adding muscles to their skeleton allowed them to move in leaps and bounds, and the benefits were felt across their entire network of products.

Walking skeleton technique varies with the system being developed. In case of a client-server system, it will be a single screen connected for navigating to database and back to screen.

In a front-end application system, it acts as a connection between the platforms, and compilation takes place for the simplest element of the language. In a transaction process, it is walking through a single transaction.

Following are the techniques that can be used to create a walking skeleton:

Methodology Shaping: Gathering information about prior experiences and using it to come up with the starter conventions. Following two steps are used in this technique:
1. Project interviews
2. Methodology shaping workshop

Reflection workshop: A particular workshop format for reflective improvement. In the reflection workshop, team members discuss what is working fine, what improvements are required, and what unique things will be added next time.

Blitz planning: Every member involved in project planning notes all the tasks on the cards, which will then be sorted, estimated, and strategized. Then the team decides on the resources such as cost and time and discusses about the road blocks.

Delphi estimation: A way to come up with a starter estimate for the total project. A group consisting of experts is formed, and opinions are gathered with an aim to come up with highly accurate estimates.

Daily stand-ups: A quick and efficient way to pass information around the team on a daily basis. It is a short meeting to discuss status, progress, and setbacks. The agenda is to keep meetings short. This meeting is to identify the progress and road blocks in the project.

Agile interaction design: A fast version of usage-centered design, where multiple short deadlines are used to deliver working software without giving important considerations to activities of designing. To simplify the user–interface test, LEET, a record/capture tool is used.

Process miniature: A learning technique as any new process is unfamiliar and time-consuming. When the process is complex, more time is required for new team members to understand how different parts of the process fit. Time taken to understand the process is reduced with use of Process Miniature.

Side-by-side programming: An alternative of pair programming is "Programming in pairs." Here two people work on one assignment by taking turns in providing input and mostly on a single workstation. It results in better productivity, and cost consumed for fixing bugs is less.

Programmers work without interfering in their individual assignments and review each other's work easily.

Burn Charts: This tool is used to estimate actual and estimated amount of work against the time.

Object-Oriented Analysis and Design (OOAD)

Object-Oriented Analysis and Design (OOAD) is an agile methodology approach toward SE and eschews traditional systems design processes. Traditional methods demand complete and accurate requirement specification before development; agile methods presume that change is unavoidable and should be embraced throughout the product development cycle. This is a foreign concept to many systems engineers that follow precise documentation habits and would require an overhaul of project management architecture in order to work. If the necessary support is in place to allow for this approach, it works by grouping data, processes, and components into similar objects. These are the logical components of the system that interact with each other.

For example, customers, suppliers, contracts, and rental agreements would be grouped in a single object together. This object would then be managed by a single person with complete executive control over the data and relationships within. This approach is people-based, relying on the individual competencies and exquisite knowledge of their respective object. The systems manager then needs to link all of the people and their objects together to create the final system. The approach hinges upon each person perfecting their object that they are in charge of, and the systems engineer puts all of the pieces together. It puts all of the design control in the hands of the individual engineers. The most popular venue for use of this type of SE is software engineering. It allows experts within their fields to focus on what they do best for a program. OOAD does not allow for convenient system oversight, process verification, or even schedule management and as such makes it very difficult to get consistent project updates. While it may be conducive to small team projects, this method is unlikely to be feasible for any Air Force project of record.

Digital Data Input–Process–Output

A systems approach to AI implementation has three sequential elements: inputs, processes, and outputs. Digital communication is the backbone controlling inputs, connecting the processes, and monitoring outputs to promote fast, responsive actions to achieve AI systems performance. The digital era has vastly improved communications in all three parts, thus speeding up the delivery of products and services tremendously, improving quality, and consequently increasing the ability to produce custom products at the same speeds and quality. If the different elements of the organization are integrated as a cohesive enterprise, communication can lead to higher efficiencies, better operational effectiveness, and cost savings. These are discussed in the framework of inputs, processes, and outputs:

Inputs: Inputs typically consist of a clear organizational vision, resources, skilled workforce, process requirements, customer desires, raw materials, market structure, and so on. Each organization must identify and define its collection of pertinent inputs, which can be variable and dynamic. Digital techniques are useful for tracking, storing, and quickly retrieving information about inputs. Data sources, digital signatures, collaboration, support, change management, organizational management system, and time-stamping of activities can be components of managing inputs in a digital environment.

Processes: Processes are the inherent capabilities of the organization to utilize the inputs to generate products, services, and/or results. Typical processes within an organization may include policies, procedures, production capabilities, design, optimization models, and so on. The digital process of an organization uses the inputs and organization strategy to transform ideas into results into outputs. Digital controls are now embedded in the processes to ensure that the end goals are achieved. Data security, credentialing, commercial external interfaces, information sharing, training, cyber security, outsourcing, cloud computing, and learning systems can be parts of an organization's processes, helping to monitor and coordinate the process for optimal effectiveness. For example, sensors embedded in equipment can sense out-of-control conditions faster than ever before and alert the operator, allowing for quick response in correcting the condition, reducing waste. Utilizing bar codes and scanners can help track raw materials, allowing for signals to be sent to the suppliers directly, reducing replenishment times and inventory levels.

Outputs: Outputs consist of physical products, desired services, and/or essential results. New inventions, a better customer experience, resilience, adaptability, concept convergence, an enhanced strategy, and so on can be parts of an organization's expected outputs. Outputs are better achieved when they are linked to the available resources and the governing processes of the organization. Signals can be sent back to production scheduling that an item has been sold, allowing for faster replenishment; shipping can be automated to pack, signal the shipper for pick up, and track delivery, speeding up the delivery and enhancing communication to the customer.

Data is the basis for systems performance. In the digital era, this is even more critical. The use of computers is the basis for all digital tools. This requires data. Every decision requires data collection, measurement, and analysis. In typical industrial engineering fashion, data may need to be collected on decision factors, costs, performance levels, outputs, and so on. In practice, we encounter different types of measurement scales depending on the particular items of interest. The different types of data measurement scales that are applicable are presented below. The different scales are important for different digital applications.

- Nominal Data Scale

  Nominal scale is the lowest level of measurement scales. It classifies items into categories. The categories are mutually exclusive and collectively exhaustive. That is, the categories do not overlap, and they cover all possible categories of the characteristics being observed. For example, in the analysis of the critical path in a project network, each job is classified as either critical or not critical. Gender, type of industry, job classification, and color are examples of measurements on a nominal scale.
- Ordinal Data Scale

  Ordinal scale is distinguished from a nominal scale by the property of order among the categories. An example is the process of prioritizing project tasks for resource allocation. We know that first is above second, but we do not know how far above. Similarly, we know that better is preferred to good, but we do not know by how much. In quality control, the ABC classification of items based on the Pareto distribution is an example of a measurement on an ordinal scale.

- Interval Data Scale

  Interval scale is distinguished from an ordinal scale by having equal intervals between the units of measurement. The assignment of priority ratings to project objectives on a scale of 0–10 is an example of a measurement on an interval scale. Even though an objective may have a priority rating of zero, it does not mean that the objective has absolutely no significance to the project team. Similarly, the scoring of zero on an examination does not imply that a student knows absolutely nothing about the materials covered by the examination. Temperature is a good example of an item that is measured on an interval scale. Even though there is a zero point on the temperature scale, it is an arbitrary relative measure. Other examples of interval scale are IQ measurements and aptitude ratings.

- Ratio Data Scale

  Ratio scale has the same properties of an interval scale, but with a true zero point. For example, an estimate of zero-time unit for the duration of a task is a ratio scale measurement. Other examples of items measured on a ratio scale are cost, time, volume, length, height, weight, and inventory level. Many of the items measured in engineering systems will be on a ratio scale.

Another important aspect of measurement involves the classification scheme used. Most systems will have both quantitative and qualitative data. Quantitative data require that we describe the characteristics of the items being studied numerically. Qualitative data, on the other hand, are associated with attributes that are not measured numerically. Most items measured on the nominal and ordinal scales will normally be classified into the qualitative data category while those measured on the interval and ratio scales will normally be classified into the quantitative data category. The implication for engineering system control is that qualitative data can lead to bias in the control mechanism because qualitative data are subject to the personal views and interpretations of the person using the data. As much as possible, data for an engineering systems control should be based on a quantitative measurement. As a summary, examples of the four types of data measurement scales are presented below:

- Nominal scale (attribute of classification): Color, Gender, Design Type
- Ordinal scale (attribute of order): First, Second, Low, High, Good, Better

- Interval scale (attribute of relative measure): Intelligence Quotient, Grade Point Average, Temperature
- Ratio (attribute of true zero): Cost, Voltage, Income, Budget

Notice that temperature is included in the "relative" category rather the "true zero" category. Even though there are zero-temperature points on the common temperature scales (i.e., Fahrenheit, Celsius, and Kelvin), those points are experimentally or theoretically established. They are not true points as one might find in a counting system.

Communication is at the core of digital implementations. Digital communication, if structured properly, facilitates enterprise transformation, which is a strategy that encompasses services and operations that directly or indirectly impact an organization's digital experience. Communication is key for success in a digital platform. Communication provides the foundation for what needs to be done, why, when, where, how, and by whom. In the digital era, the ubiquitous availability of new communication modes, tools, and timing has enhanced communication, which, in turn, has sped up commerce. Speed is expected today—next day and now same day delivery is becoming common place, and having results, knowing what is going on, on the production floor, in supply chains, and about financial transactions all now require fast communication. It is essential to leverage the new communication assets to achieve the desired performance level.

Technology has changed communication in positive ways. However, there can be pitfalls in the speed and reach of digital communication. Safeguards must be put in place to ensure preservation of the integrity of communication and the information it conveys. Today's fast-paced environment requires close, careful communication. No longer can we rely on walking over to another office or traveling to meetings to discuss issues, concepts, and operations. We cannot only view the immediate environment—we must understand the bigger system and our place in it, particularly with a digital footprint.

Cooperation and coordination are paramount to a successful implementation of any digital improvement program. Spending more does not always translate to better outcomes. The trade-offs between time, budget, and performance must chart a path through the application of digital tools and techniques. The current era of digital operations has created better communication to improve resource utilization by providing the ability to collaborate virtually, reducing the resources needed to travel for meetings, and giving more time to personnel for value-added efforts. However, this increased speed of communication can also make the waste and ineptitude worse, faster. Some of the IE tools for digital design and analysis for improving digital systems communication include operations research, modeling and simulation, AI, big data analysis, digitally networking assets, and supply chain management. Companies and industries that in the past have not embraced AI strategies are now, with the help of digital tools, pursuing it.

DIGITAL COLLABORATIONS

Not all AI pursuits are technical or mechanical. The human elements involved in implementing AI require that we also consider the soft side of AI project management. A natural progression in AI implementations is to leverage the capabilities of other organizations through strategic alliances, rather than developing duplicate capabilities in house. Strategic alliance is defined as a formal alliance or "joining of forces" between two or more independent organizations for the purpose of meeting mutual business goals. Each partner in the alliance has something to bring to the "table," such as products, supply chain, distribution network, manufacturing capability, funding, capital equipment, operational expertise, know-how, or intellectual property. Strategic partnering represents cooperation whereby project management synergy ensures that each partner derives benefits beyond normal independent operation.

While there are pros and cons to partnering, the advantages often outweigh the downsides. Advantages of strategic partnering include the following:

1. It allows each partner to concentrate on operations that best match its capabilities.
2. It permits partners to learn from one another and develop competencies that may be readily utilized elsewhere.
3. It facilitates synergy that increases the outputs of both partners' resources and competencies.

Today's world requires better utilization of limited resources. More and more, cooperative partnerships are needed to achieve the highest effective use of these resources. Cooperation is a basic requirement for resource interaction and integration in any partnership. Digital communication is immediate and goes globally, sending the exact same information to all involved, a key requirement for the best cooperation. More projects fail due to a lack of cooperation and commitment than any other project factor. This lack often happens when some of the needed people are not informed in time or get different communications. To secure and retain the cooperation of partners, the most positive aspects of a proposed partnership should be the first items of communication. Such structural communication can pave the way for acceptance of the proposal and subsequent cooperation. Then the other aspects such as profit sharing and finances, staffing, facility use, performance reporting, conflict resolution can be negotiated.

There are different types of cooperation, as summarized below:

- *Functional Cooperation*: This is cooperation induced by the nature of the functional relationship between two partners. The two partners may be required to perform related functions that can only be accomplished through mutual cooperation.
- *Socially Responsible Cooperation*: This is the type of cooperation effected by a socially responsible relationship between two partners. This is particularly common for activities that may impact the environment. The socially responsible relationship motivates cooperation that may be useful in executing entrepreneurial partnerships.
- *Regulatory Cooperation*: This is usually cooperation that is based on regulatory requirements. It is often imposed through some legal authority and expectations. In this case, the participants may have no choice other than to cooperate.
- *Industry Cooperation*: This is cooperation that is fueled by the need to comply with industry standards and build a consensus to advance the overall industry in which partners find themselves. For example, in the early days of cell phones, the Cellular Trade Industry Association (CTIA) developed market alliances to refute the early fears about the health impacts of using cell phones. The group sponsored several research studies at universities to confirm that cell phones were safe. Such a monumental undertaking could not have happened without market cooperation. However, caution must be exercised to ensure that cooperation does not degenerate into market collusion, which is illegal.
- *Market Cooperation*: In order for each player in the market to thrive, the overall market must be vibrant. For this reason, market cooperation involves partnering of market players to advance market vitality. This usually happens in evolving markets. While similar to industry cooperation, this is focused in specific market areas. For example, regions will set up local market associations to protect certain foods so that only that region can produce the food. These associations cooperate on geographical or origin designation or manufacturing process to protect the local market against the wider food industry.
- *Administrative Cooperation*: This is cooperation brought on by administrative requirements that make it imperative that two partners

work together toward a common goal, such as market growth. In fact, market cooperation and administrative cooperation can coexist across organizations. One good example of administrative cooperation is the monthly meeting partnering between two professional associations in the same local community. The authors have participated in cohosted meetings between IISE (Institute of Industrial and Systems Engineers) and ASQ (American Society for Quality).

- *Associative Cooperation*: This is a type of cooperation that may be induced by collegiality. The level of cooperation is determined by the prevailing association that exists between the partners. Industry associations often cooperate under this approach.
- *Proximity Cooperation*: This type of cooperation may be viewed as "Silicon-valley orientation" whereby organizations located within the same geographical setting form cooperating alliances to pursue mutual market interests. In an ideal case, being close geographically should make it possible for partners to work together. In cases where the ideal expectation does not materialize, explicit efforts must be made to encourage cooperation.
- *Dependency Cooperation*: This is cooperation caused by the fact that one partner depends on another partner for some important aspect of its operation and business survival. Such dependency is usually of a mutual two-way structure. Each partner depends on the other partner for different things.
- *Imposed Cooperation*: In this type of cooperation, external forces are used to induce cooperation between partners. This is often the case with legally binding requirements.
- *Natural Cooperation*: This is applicable for cases where the two partners have no way out of cooperating. Physical survival requirements often dictate this type of cooperation.
- *Lateral Cooperation*: Lateral cooperation involves cooperation with peers and immediate contemporaries in the market place. Lateral cooperation is often possible because lateral relationships create an environment that is conducive for mutual exchanges of information and operational practices. An example is the recent bailout pursuit by the big three of the US auto industry.
- *Vertical Cooperation*: Vertical or hierarchical cooperation refers to cooperation that is implied by the hierarchical structure of the market in which the partners operate. For example, subsidiaries are expected to cooperate with their vertical parent organizations.

Whichever type of cooperation is available or needed in a situation, the cooperative forces should be channeled toward achieving mutual goals in the most effective way. Documentation of the level of cooperation needed will clarify roles, responsibilities and boundaries and win further support and sustain joint pursuits. Clarification of organizational priorities will facilitate personnel cooperation. Relative priorities of multiple partners should be specified so that a venture that is of high priority to one segment of the partnership engagement is also of high priority to all partners within the endeavor.

LEAN AND SIX SIGMA IN AI

Lean or continuous process improvement is an ongoing systematic effort to improve day-to-day operations to remain productive and operationally efficient. For example, AFSO21 is a coordinated pursuit of operational improvements throughout the Air Force by way of an integrated process of applications of various tools and approaches for improving operations while reducing resource expenditures. Productivity, quality of service, process enhancement, flexibility, adaptability, work design, schedule optimization, and cost containment are within the scope of AFSO21. Increases in operational efficiency are best accomplished through gradual and consistent closing of gaps rather than the pursuit of one giant improvement step. The practice of drastic or sudden improvement often impedes process optimization goals.

Lean is a way of doing business, a culture that focuses the entire organization on the identification and elimination of sources of waste in operations. By comparison, the Six Sigma program uses a subset of Lean tools to identify and eliminate the sources of defects through the reduction of variability, a type of waste. When Lean and Six Sigma are combined, an organization can reduce both waste and defects in operations. Consequently, the organization can achieve higher performance, better employee morale, more satisfied constituents, and more effective utilization of limited resources.

The basic principle of Lean is to take a close look at the elemental composition of a process to eliminate non-value-adding elements or waste. Lean and Six Sigma techniques use analytical and statistical techniques as the basis for pursuing improvement objectives. But the achievement of those goals depends on having a structured approach to the activities associated with what needs to be done. If an IE approach is embraced at the outset, it will pave the way for achieving Six Sigma results and make it possible to realize Lean outcomes. The key in any operational management endeavor is to have a structured plan so that diagnostic and corrective steps can be pursued.

If inefficiency is allowed to creep into operations, it would take much more time, effort, and cost to achieve a Lean Six Sigma cleanup. To put the above concepts in a military perspective, Six Sigma implies executing command and control processes such that errors are minimized in the long run. Likewise, the techniques of Lean ensure that only value-adding command and control actions are undertaken. This means the elimination of waste. This brings to mind Parkinson's Law of bureaucracy, which states that "work expands to fill the time available," as a result of which unnecessary activities are performed. Military leaders must ensure that functions do not extend needlessly just to use up available time and resources. Short and effective functions are better than protracted ones that result in counterproductive results. It is of note to mention that the popular Murphy's Law, which states that "whatever will go wrong will go wrong," provides a cautionary note about overreliance on digital operations. As we often experience in the virtual environment created by the COVID-19 pandemic, whatever can go wrong in the digital virtual environment has been known to go wrong. For a digital systems framework, careful planning, precautions, and contingency actions can help preempt operational failures. Below is a summary of the common rules, laws, and principles encountered in an operational setting:

Parkinson's law: Work expands to fill the available time or space.
Peter's principle: People rise to the level of their incompetence.
Murphy's law: Whatever can go wrong will go wrong.
Badiru's rule: The grass is always greener where you most need it to be dead.

Humans in the Loop in AI Implementations

Humans must remain in the loop of AI. Based on the examples presented in Chapter 1, human intuition is still needed in most automated systems. The attributes and capabilities of humans in the AI loop must be taken into consideration when designing AI systems. Person-to-person communication can offer a backup or recovery safeguard when automated systems fail or diminish in performance. Human task assignments can supplement machine assignments. An assignment to a task implies being involved in various aspects of the task's requirements. Communication complexity is one aspect that is often overlooked in the assignment of personnel to tasks where they do not add any value. The increased requirement for multi-person and hierarchical communication is the subtlest of wastes in any organization because it not only impedes overall organizational performance, but it also reduces the utilization of critical resources. A structural review of communication, cooperation, and coordination of personnel can identify where resource utilization improvement can be achieved. Often the assignment requirements within the Air Force are not in sync with the prevailing constraints. The resolution of such problems requires integrative communication, cooperation, and coordination. This is achieved through a systems-based approach. Basic questions of what, who, why, how, where, and when should always be addressed strategically when doing personnel and resource assignments. It highlights what must be done and when, with respect to the following:

- Does each participant know what the objective is?
- Does each participant know his or her role in achieving the objective?
- What obstacles may prevent a participant from playing his or her role effectively?

Communication complexity increases with an increase in the number of decision points. It is one thing to wish to communicate broadly, but it is another thing to create complexity when more decision points are involved. For example, the military structure typically involves multilayer decision processes. While this provides for more chains of command with embedded checks and balances, it can impede overall systems efficiency by creating additional layers of decision. The versatility, ease, and flexibility of the digital era may lure us into a situation of complex communication channels that impact complexity on operations to the point of diminishing value rather than improving performance. As the number of those involved increases, so increases the complexity of the operations.

CONCLUSION

The work environment must be designed to facilitate cooperation in support of enterprise transformation programs. Team cooperation is influenced by proximity, functional relationships, professional affiliations, social relationships, official capacity influence, formal authority influence, hierarchical relationships, lateral camaraderie, intimidation, and enticement. Operational coordination can be achieved through teaming, delegation, empathetic supervision, partnership, token-passing, and deliberate operational baton hand-off. These are all systems-oriented processes that should be leveraged for digital enterprise transformation. Communication complexity can impede the overall performance of an organization. In the analogy to computational science, communication complexity studies the amount of communication required to solve a problem when the input to the problem is distributed among two or more parties. Communication complexity can be approximated as the number of potential unique conversations that can occur in a group. Below are some factors that could affect communication complexity:

- Interpersonal conflicts
- Office politics
- Leadership style
- Communication tools
- Plainness of the communication content
- Analogy versus digital communication
- Traceability of communication contents
- Message framing and context

Using digital communication tools can help reduce complexity, conflicts, and politics, enhance leadership, and speed up communications. Choosing the right tool, and designing the use of the tool, is key to successful communication. In this regard, the emerging focus on 5G technology should be embraced. 5G is the fifth generation of mobile technology, which is an advancement beyond the present 4G technology. 5G provides the platform for new and emerging technologies, such as Internet of Things (IoT), AI, and Big Data, to improve the way we live, work, and pursue better systems performance.

REFERENCES

Badiru, A. B. (2014a), "Quality insights: The DEJI model for quality design, evaluation, justification, and integration," *International Journal of Quality Engineering and Technology*, Vol. 4, No. 4, pp. 369–378.

Badiru, A. B. (2019), *Systems Engineering Models: Theory, Methods, and Applications*, Taylor & Francis Group/CRC Press, Boca Raton, FL.

Badiru, A. B., editor (2014b), *Handbook of Industrial & Systems Engineering*, Second Edition, Taylor & Francis Group/CRC Press, Boca Raton, FL.

Neural Networks for Artificial Intelligence

4

INTRODUCTION

Neural networks are recognized as one of the basic foundations for the simulated intelligence in artificial intelligence systems. This chapter presents narrative introduction to neural networks. The typical complex mathematical representations and schematics of neural networks are not included in the chapter since this is a condensed focused book. Readers interested in the mathematical and graphical exposition of neural networks may refer to Badiru and Cheung (2002), Badiru and Sieger (1998), Sieger and Badiru (1993), and Milatovic et al. (2000).

The whole basis for neural networks is to train a network of electronic connections to mimic intelligent connections of human neurons so that the knowledge gained can be used for similar decision problem scenarios. In definitional terms, a neural network is a network or circuit of neurons or artificial neural network, composed of artificial neurons or nodes. A neural network is either a biological neural network, made up of real biological neurons, or an artificial neural network, for solving artificial intelligence problems. The connections of the biological neuron are modeled as weights. A positive weight reflects an excitatory connection, while negative values mean inhibitory connections. All inputs are modified by a weight and summed. This activity is referred to as a linear combination. An activation function is used to control

DOI: 10.1201/9781003089643-4

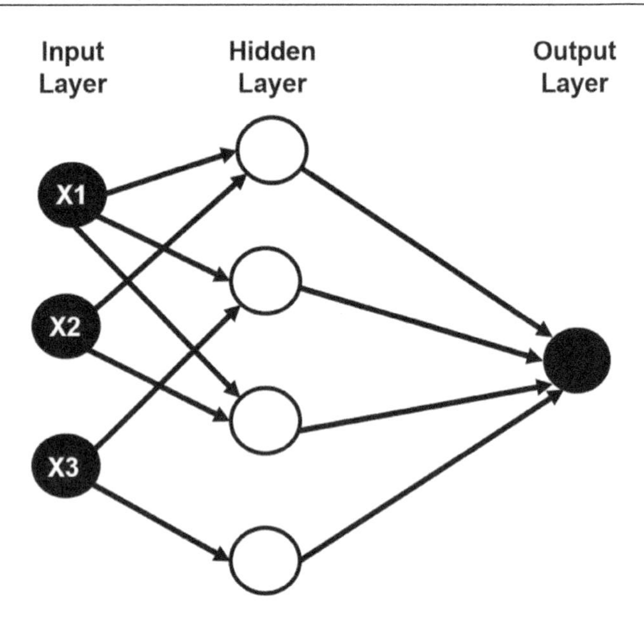

FIGURE 4.1 Basic layout of a simple feedforward neural network.

the amplitude of the output. An acceptable range of output is usually between 0 and 1, or it could be –1 and 1. Figure 4.1 shows a basic layout for a simple feedforward neural network. The network gets increasingly complex as more inputs nodes and weights are added to the problem scenario.

Many complex systems are built from simple foundational elements. The field of Boolean algebra came into existence due to the selection of simple elements that can only take on two values: 0 and 1. These elements are combined by three basic operations: AND, OR, and NOT. From these simple elements and operations, complex combinational circuits and sequential circuits can be constructed, eventually culminating in today's computers. Mathematicians also have long worked with set theory, an extension where a set may consist of more than 0 and 1 values. The Boolean operations AND, OR, and NOT have been extended for set theory to UNION, INTERSECTION, and COMPLEMENT. More recently, the concept of fuzzy logic has been introduced by Zadeh. In fuzzy logic, the participation of an element's membership in a set is not binary, i.e., none or all, but rather in degrees between none and all.

In the field of signal processing, many modeling techniques are based on a simple linear model: the linear combiner where the output is a linear combination of the weighted inputs. Much study has been devoted to finding the best techniques for estimating the parameters of a linear model. Statistical techniques have long been developed for linear regression that is a form of a linear combiner. Deterministic signal processing techniques for linear models were well developed in recent decades. Adaptive signal processing techniques provide a means to iteratively obtain model parameters, thus simplifying the application of many computationally intensive approaches.

The original motivation for studying neural networks stems from the desire to perform by machines complex and intelligent tasks that only the human brain can do. It is commonly accepted that while the response time of a nerve cell is in the order of milliseconds, the collective intelligence of billions of billions of nerve cells in the human brain is staggering. A human brain is able to recall images and events that occurred decades ago. The brain is also able to recognize a single face out of the millions of faces and images seen before. Yet the study of anatomy has shown that on the surface, the "operation" of a single brain cell is extremely limited. It is indeed intriguing to see that the totality of these simple nerve cells can achieve the wonders of modern technology. Hence, there is tremendous interest by researchers in recent decades in studying the biological neural networks in an attempt to construct artificial neural networks from simple artificial neurons that would equally achieve what two and half pounds of a typical human brain can attain.

DEFINITION OF A NEURODE

A biological nerve cell is composed of a cell body. A large number of tree-like incoming branches called dendrites collect sensory inputs to the cell body. The dendrites make "contact" with other nerve cells through a synapse. When a nerve cell is put into an active state, it fires i.e., an electrical impulse is generated and sent along a conduit called an axon. When the electrical impulse passes through a synapse, a number of chemical reactions are evoked causing specific chemicals to be released and collected by the dendrites. When enough sensory inputs have been collected by the dendrites of the receiving nerve cell, it in turn fires. The efficacy of the electrical stimulation from an impulse on an axon passing through a synapse is variable dependent on a large number of parameters such as the availability of ions and the general health of the body.

In a typical neurode diagram, a set of inputs, each with its own associate weight, feeds into a neural function, which operates like a process, to generate an output. The function is a mathematical operation that converts the weighted inputs into a coordinated output. An artificial neuron or neurode is a simple processing element patterned after the biological model. A neurode is composed of a number of inputs and a single output. Each synapse of a biological neuron represents an input of the neurode. The efficacy of the synaptic junction is represented by a multiplicative weight associated with that neurode input. The coincidence of the synaptic stimulation is modeled by a simple summation. The activation of the neurode is defined by an activation of the weighted sum of the input. In equation format, the output, o, of a neurode can be expressed as an output of a function, where x_i are the inputs, w_i are the weights associated with the inputs, and f is a nonlinear function. There are a total of N inputs. The same equation can also be expressed in matrix form. In general terms, the constant threshold of a neural network is defined as x_0, and it is set to 1.

The mathematical definition immediately highlights the similarity between a neurode and a linear combiner. In fact, the net parameter is precisely a linear combiner. The primary difference between a neurode and a linear combiner is that for a neurode, there is an additional nonlinear transformation, f, following the linear combiner. In most cases, this transformation or activation function is taken to be nonlinear. In the extreme case when the activation is taken as a linear function and more specifically a linear ramp function, then a neurode is the same as a linear combiner. In other words, a neurode is a linear combiner and more.

VARIATIONS OF A NEURODE

There are a number of variations on the format of a neurode. Most of the variations stem from the different definitions of the input range for x and in the choice of the activation functions f. The weights of the neurode are usually represented as real numbers. If the activation function produces only discrete values due to a threshold function, the output is said to be discrete. Most other activation functions produce continuous outputs. The activation can be a simple linear function, though most of the time, the activation function is taken to be a nonlinear function. The following lists a number of commonly used activation functions.

- Discrete output
  - Unipolar outputs
    - Threshold function (or hard limiter or Heaviside function):
    - Stochastic function:
  - Bipolar outputs
    - Sign(net):
    - Stochastic function:

Continuous output
- Unipolar outputs
  - Linear function
  - Piecewise function
  - Sigmoidal function
- Bipolar outputs
  - Signum function
  - Linear function
  - Piecewise function
  - Hyperbolic tangent function

The use of different activation functions produces different results. While there have been reports of special activation functions with special characteristics, e.g., periodic, most activation functions are monotonically increasing. The hyperbolic tangent function is of particular interest because the steepness of the curve can be adjusted with different values of the λ parameter. When λ increases, the transition becomes steep. It is worthwhile to note that when λ approaches infinity, the hyperbolic tangent function approaches the sign function. The same observation can be said related to the sigmoidal function. As the λ parameter of the sigmoidal function approaches infinity, the sigmoidal function approaches the hard limiter function.

SINGLE NEURODE: THE MCCULLOUGH-PITTS NEURODE

One of the early studies of the behavior of a neurode is the McCulloch–Pitts neurode (see Badiru and Cheung, 2002). The input is discrete $x \in \{0,1\}$ and the output is also discrete $o \in \{0,1\}$. The activation function is a simple threshold function.

There are many uses for a McCulloch–Pitts neurode. One notable observation is that one or more neurodes can be arranged to function as simple Boolean elements: an AND, OR, or NOT gate. For example, if a McCulloch–Pitts neurode takes on a single input, and let the weights $W=[-1.001,1]$, then the neurode functions as a NOT gate. However, given a two-input McCulloch–Pitts neurode, if $W=[0.001,1,1]$, then the neurode functions as an OR gate. If $W=[1,1,1]$, then the neurode functions as an AND gate. This implies that all combinational circuits composed of AND, OR, and NOT components can be directly replaced by McCulloch–Pitts neurodes and hence a neural network.

Using the basic principle of a McCulloch–Pitts neurode, more complex and useful structures can be built. For example, McCulloch–Pitts neurodes can be connected to produce an analog–digital converter using the successive comparison approach. The single input x is a continuous parameter between 0 and 1. The output o_1–o_4 represents four binary bits. Each of the binary bits is produced by a single McCulloch–Pitts neurode. The first neurode compares with half of the dynamic range and reports whether the input x is above or below the half-way mark. This determines the most significant bit. If this bit is 1, then half of the dynamic range is subtracted from the input, otherwise the input is left alone. The comparison continues in like manner with the next significant bit until the desired resolution is reached.

SINGLE NEURODE AS BINARY CLASSIFIER

A McCulloch-Pitts neurode can also be used as a binary classifier that separates the input into two classes. This is done by examining or unraveling the equation for the output. Assume that the neurode has two inputs x_1 and x_2.

Clearly, the output separates the entire input space into two classes represented by $o=1$ and $o=0$. The decision boundary is the hyperline that separates the input space into these two classes.

The input space on one side of the hyperline belongs to one class while the input space on the other side belongs to the other class. Hence a single neurode is a binary classifier.

Given two clusters of raw data, it is of interest to determine the parameters of the binary classifier, i.e., determine $W=[w_0, w_1, w_2]$. If the locations of the clusters are known, then the hyperline can be determined immediately by realizing that the equation of a line can also be written using the segment bisector form, where a and b are the locations of the clusters, and ||.|| represents the Euclidean norm of the enclosed vector. The same technique can be derived using the principle of minimum distance. Hence the above equation is also called the minimum distance classifier. If the locations of the two clusters are not known and only the original data points are given, then an iterative technique can be applied to determine the weights of the neurode.

A special case of the general binary classifier is the Bayesian classifier where the decision boundary is drawn according to the maximum likelihood function. Assume that the elements of the two classes have a Gaussian distribution with different means (μ_1 and μ_2). Let C be the covariance matrix of the combined distribution.

SINGLE NEURODE PERCEPTRON

In general, the weights associated with a neurode with an arbitrarily chosen activation function can be determined, albeit iteratively. In some literature, a neurode is also called a perceptron. A single discrete perceptron can be trained to produce a particular desired output in response to a particular input. The goal is to determine the weights W such that the predicted output o is the same as the desired output d. The problem can be solved iteratively by minimizing the total prediction error.

The same procedure can also be applied to a continuous perceptron, i.e., a perceptron whose output is a real number. The problem can be restated as follows. Given a set of input–output pairs, the goal is to determine the weights W such that the predicted output o is the same as the desired output d. The problem can be solved iteratively by minimizing the total prediction error. The factor f'(net) is the derivative of the activation function. Note that the adaptation equation includes this additional term because of the relationship between the neurode output and the activation function. The adaptation equation is repeatedly applied for all input data points until the error is below the acceptable threshold.

Single-Layer Feedforward Network: Multi-Category SLP

A single neurode or perceptron can function only as a dichotomizer partitioning the input space into two classes. Hence, more than one perceptron is needed when there are more than two classes. This is called a multi-category single-layer perceptron (SLP) or simply SLP. Putting together more than one perceptron results in a single-layer feedforward network. Since the output of each perceptron is independent of the others, the training procedures for a single discrete or continuous perceptron can be readily extended to the single-layer feedforward case.

ASSOCIATIVE MEMORY

Looking from a broader perspective, an SLP provides a transformation from a multidimensional input to a multidimensional output. In other words, an SLP provides a mapping from input to output such that f: $R^N \rightarrow R^M$ where N is the input dimension and M is the output dimension. This is commonly called an associative memory.

A linear associator is obtained when the activation function is taken to be a ramp function, i.e., $f(z)=z$. In this case, the output is simply $O=W^TX$, i.e., the output is a linear combination of the input, hence the name. A linear associator is often used to map the input space to the output space. Given a set of input–output pairs, where K is the number of input–output pairs, X_i represents the ith pair of the given input vector, and D_i represents the corresponding output vector, the problem is to determine the weight matrix, such that the output of the linear associator is the same as the desired output.

CORRELATION MATRIX MEMORY

Hebbian learning (also called the correlation matrix memory) is an approach for determining the weights of a linear associator based on the principle of classical conditioning proposed by Hebb. In classical conditioning, an unconditioned stimulus (food) causes an unconditioned response (salivation) in a system (an animal). A conditioned stimulus (bell) originally causes no specific response. By pairing the unconditioned stimulus and the conditioned stimulus together, an association is made in the system such that after a while, the conditioned stimulus also elicits a conditioned response similar to the unconditioned response. Hebb postulated that the unconditioned response (neuronal output) causes an association (increases weights) to be formed with the input (unconditioned stimulus AND conditioned stimulus that were presented simultaneously). In other words, the excited output of a neurode strengthens the associated weights of the corresponding excited output. Mathematically speaking, this association can be written as the outer product of the input and output vector. The correlation matrix approach is simple to apply, but perfect recall dictates that the input vectors must be orthonormal.

Another way to look at the problem is from the numerical analysis perspective. All the supplied input–output pairs can be written as a system of linear equations. There are KxM equations represented by the above matrix equations, and there are $N \times M$ unknowns in the weight matrix. As long as K is larger than N, we have an overdetermined system of linear equations. In most cases, the number of given patterns is usually much greater than the dimension of the input vector. The application of the least mean squares (LMS) approach leads to the pseudo-inverse solution of the above equation. Hence, this is called the pseudo-inverse approach. For this approach, define X as the input matrix and D as the desired output matrix. X and D are now in matrix form containing all the given data.

The pseudo-inverse approach is also a single-step training procedure. While the approach seeks to minimize the total mean square error, it does not allow the ability to fine-tune the derived weight matrix.

WIDROW–HOFF APPROACH

While the pseudo-inverse approach is based on the mean-square-error principle, another approach is based on the least-expected-error principle commonly known as the Widrow–Hoff equations.

This approach is attractive because the weight matrix can be obtained in one step similar to some of the other approaches discussed so far. Statistically speaking, the weight matrix thus obtained is optimal.

LMS APPROACH

The LMS algorithm of Widrow and Hoff is the world's most widely used adaptive algorithm, fundamental in the fields of signal processing, control systems, pattern recognition, and artificial neural networks. These are very different learning paradigms.

When the dimension of the input and output vectors becomes large, the dimension of the weight matrix also becomes correspondingly large. It becomes impractical to apply the Widrow–Hoff approach because of the matrix inversion operation. Widrow has also developed an iterative method to adaptively determine the weight matrix using the principle of steepest descent. Instead of using the expected error, the algorithm uses the instantaneous error.

The LMS algorithm is attractive because it is simple to implement and is iterative. In some neural network literature, this approach is also known as the delta rule.

ADAPTIVE CORRELATION MATRIX MEMORY

Previously, it has been shown that the correlation matrix memory approach is a single-step training, single-step recall method and hence does not allow for fine-tuning. This approach can be modified to so that the weights are adaptively obtained. The new approach is called the adaptive correlation matrix approach. The initial weights are set to zero at the beginning. Subsequent presentation of the input sample points causes the weights to adaptively settle on the best value for predicting the output. This adaptive modification provides the ability for the algorithm to fine-tune and obtain the best values for the weights according to the principle of Hebbian learning.

ERROR-CORRECTING PSEUDO-INVERSE METHOD

The same approach can also be applied to the pseudo-inverse approach. Previously, it has been shown that the correlation matrix memory approach is a single-step training, single-step recall method and hence does not allow for fine-tuning. This approach has been further modified to allow fine-tuning by adding an adaptive stage at the end. The new approach is called the error-correcting (or iterative) correlation matrix approach, after the initial weights are determined according to the correlation matrix approach. The weight matrix is further refined by iteratively correcting the parameters. The error-correcting pseudo-inverse matrix approach allows the user a way to fine-tune the weights so that the prediction can more closely approach the desired values.

SELF-ORGANIZING NETWORKS

In many practical applications, it is not unusual that the dimension of the input space is large. Some parameters may be significant while others may not. It is always of interest to determine which parameters play a more significant role than others in the input–output relationship. It is further of interest to find out if the given sample points can be modeled by a smaller number of parameters. In other words, the original space is called the data space and may be of high dimension composed of all the input parameters. We would like to find a mapping that would transform the sample points from the data space to the feature space. The feature space contains a small number of parameters, yet still contains all the essential characteristics originally contained in the data space. This is called data reduction.

Classification can be considered as data reduction. The original sample points in the data space are mapped into the feature space composed of the different classes. Regardless of the input dimension, the feature space is merely the class of the input space. The input space has been partitioned into distinct region where each region represents one class. By specifying a data point to be in a particular class means that the data point possesses all the inherent characteristics related to that class as indicated by the centroid of the class. Deviations to the centroid are considered to be random perturbations and insignificant.

Modeling can also be thought of as data reduction. Given a set of sample points, the original data space is transformed into a set of parameters related to the underlying model. The parameters of the model summarize the characteristics of the underlying model dynamics. This approach is generally called parametric because based on the specified functional form, the given sample points are modeled by specific values of the model parameters. In other words, the feature space is the collection of parameters. Parametric modeling has been widely used in many applications. The caveat in parametric modeling is that the specified functional form adequately describes the underlying model characteristics. For example, one can fit a straight line to data derived from a quadratic function. Likewise, the same data can also be fitted to a cubic. In general, the underlying model is unknown. The determination of the best model is a separate and nontrivial task in itself.

A more general approach than parametric modeling is nonparametric modeling. In the latter case, no assumptions have been made regarding the functional form. An example of nonparametric modeling is the principal components approach.

PRINCIPAL COMPONENTS

Given a set of input sample points $x_i \in R^N$, the idea is to find a small set of exemplars that can be used to collectively describe most of the sample points. There are many numerical methods that can be used for this purpose. On such technique is the eigenvalue/eigenvector analysis. The decomposition procedure is well understood and can be found in many numerical analysis texts.

The diagonal elements of the matrix are called eigenvalues. The eigenvectors are unit vectors that identify the basic characteristics of the input data in a nonparametric format. For this reason, the eigenvectors are also called basis vectors.

In the above formulation, the eigenvectors form the basis of the feature space, and the eigenvalues show the weightings of the individual contributions of the eigenvectors. Since the eigenvectors are unit vectors, their contribution is completely normalized. However, the relative importance of the eigenvectors is indicated by the corresponding eigenvalues. The larger the eigenvalue, the more the associated eigenvector contributes to explaining the variance of the input data. On the other hand, if the eigenvalue is small, then the corresponding eigenvector most likely models the random perturbations that exist in the input data.

Data reduction is achieved by realizing that not all the eigenvectors are needed in reconstructing the input data matrix. Only the significant eigenvectors with large eigenvalues are used. The significant eigenvalues and their associated eigenvectors are called principal components. Since each eigenvector is orthogonal to all the others, the principal components are independent of one another. Hence, the principal components show the degree of independent processes that exist in the underlying model dynamics.

CLUSTERING BY HEBBIAN LEARNING

Suppose a set of input sample points are given. Note that no output values are given or needed at this point. It is of interest to define a model that would adequately describe the input data space. This is in essence the clustering problem. Without explicitly indicating which sample point belongs to which class, the problem at hand is to find out the number of clusters and the corresponding centroid locations of the clusters.

In the study of linear associators, the method of adaptive Hebbian learning has been presented as an effective way to determine the weights in an iterative manner. Consider a single neurode with input vector x_t, scalar output y, and time index t. Repeat the update equation for the weight vector W_t at time index t for the adaptive correlational matrix memory approach for a single neurode. It is reasoned that the same process can be used to form clusters. This is done by replacing the desired output with the actual neurode output. Beginning with initial random weights, the update equations iteratively enforce those weights that produce an output.

One major problem with the above update equation is that the weights can grow unchecked and unbounded. The more the input patterns are presented, the more the outer product adds to the weights. For this reason, this approach has not been in widespread use.

CLUSTERING BY OJA'S NORMALIZATION

In an effort to curb the unboundedness of the weights, Oja (see Badiru & Cheung, 2002) proposed to normalize the weights after each weight update. In other words, the relative importance of each weight is redistributed among all the weights. The summation term gives the power (squared value) of all the weights. The weights are therefore normalized by the square root of the total weight power. This in essence causes the weights to be bounded in magnitude.

A first-order approximation of Oja's normalization is obtained by replacing the square root operation with the first-order series expression for a square root. In the above equation, there is a positive feedback to increase the weights for self-amplification. However, at the same time, there is also a negative feedback to control the growth of the weights.

The abovementioned Oja's extension to the weight update equation results in some very interesting properties. Taking expectations on both sides of the equation, convergence is obtained when the update is zero. When convergence condition is reached, it can be shown that the weights converge to the largest eigenvector of data correlation matrix. The other eigenvector can be obtained by Hotelling's deflation. By deleting the contribution of the most significant eigenvector, a new data set can be formed so that the same technique can again be applied, thus causing the algorithm to subsequently converge to the next largest eigenvector. If the first neurode has already converged to the largest eigenvector, then based on the Hotelling's deflation principle, a new data set can be formed by subtracting the effects of this eigenvector.

Oja's extension shows that a two-layer network can be used to automatically classify the input data space into distinct clusters merely by presentation of the input patterns themselves. This is done by the addition of lateral connections at the output layer. In addition, for the algorithm to work, the convergence of the second and subsequent neurodes should be withheld until the first neurode has converged. Then the weights for the second neurode are allowed to converge and likewise for the subsequent neurodes in order.

COMPETITIVE LEARNING NETWORK

The architecture for a competitive learning network is similar to an SLP. There are no explicit connections between the neurodes at the output layer. However, the output of all the neurodes at the output layer must be considered together during the weight update process.

The operation of a competitive learning network is essentially the same as an SLP with one important distinction: not all the weights are allowed to be updated. When a pattern is presented to the network, all output neurodes examined and processed the input pattern. Each output neurode generates an output. All neurode outputs are then compared, and a winner is selected based on the largest neurode output. The winning neurode is then allowed to update its weights. The weights of all other neurodes remain the same. Hence, this is called a winner-takes-all strategy. Only the winning neurode has the privilege to be updated. The competitive learning network is also called a Kohonen network.

The update equation causes the weight of the winning neurode to be more and more like the input pattern. If there are more than one sample point that are selected by the neurode, the weights of the neurode tend to settle on the centroid of the sample points. In other words, the weights of the competitive networks yield the centroid location of the clusters. Each active neurode depicts a cluster and serves to represent the location of the associated cluster.

In summary, self-organizing networks such as the PCA networks or the competitive learning networks are SLP with special training procedures. Because of the special process involved, these networks have been shown to be useful in automatically discovering clusters in the data space.

MULTIPLE-LAYER FEEDFORWARD NETWORK

Multiple-Layer Perceptron

While a single-layer feedforward network is able to map a multidimensional input space to a multidimensional output space, each output is basically independent of one another. A multiple-layer perceptron (MLP) is formed by putting more than one layer together. An MLP is particularly attractive because the additional layers allow the results of one layer to be further processed, arranged, and put together to make a complex system.

Recall that a single neurode basically creates a hyperline in the input space. An SLP therefore is equivalent to a set of hyperlines placed in the same input space working in parallel, each hyperline for a corresponding output neurode. In order to correlate the different hyperlines together, thus creating multiple intersecting regions, additional layers are required. The first layer creates a set of hyperlines, the second layer relates the hyperlines together to form a contiguous hyper-regions. Since there can be any number of hyperlines, it is therefore possible to approximate the shape of any hyper-regions in a piecewise manner.

While two layers are needed to join separate hyperlines together to form a hyper-region, another layer is needed to join multiple disparate hyper-regions together into a single class. The third layer allows multiple hyper-regions distributed anywhere in the input space to be related together. Hence, it is commonly accepted that a three-layer feedforward neural network is capable to realize any function for this reason.

XOR Example

As an example, consider a two-layer network for the Exclusive-OR (XOR) function. An XOR function has two inputs, x and y, and a single output. The inputs and the outputs are discrete taking on values of 0 and 1. When the inputs are distinct, the output is 1. When the input is the same, the output is 0. The XOR function can be solved using two McCulloch–Pitts neurodes in the first layer and a single McCulloch–Pitts neurode in the output. There are two input neurodes because there are two inputs. A single output neurode suffices in the second layer since there is only a single output variable. Each of the two neurodes in the first layer produces a line in the input space as shown. Note that the positive decision region for the first neurode is to the left of the line while that for the second neurode is to the right of the line. By so doing, the inputs for <x, y>=<0,0> and <1,1> are left in the same region, hence producing the same output results.

Back-Error Propagation

While the MLP has the potential to approximate any function, the network can only be practically used if there exists a way to determine the weights that would approximate the desired function. Given a set of input–output pairs, it is highly desirable to be able to determine the weights directly from the given data. This problem has been solved using the generalized delta rule. The use of generalized delta rule has greatly enhanced the usage of neural networks, and many applications have been reported in the literature on how the generalized delta rule can be applied successfully to solve many practical problems.

The basic idea of training a multiple-layer feedforward neural network is to generalize the delta rule for each weight. The delta rule requires the computation of the derivative of the error with respect to the weight of interest. This can be done by repeated applying the chain rule. In some literature, the generalized delta rule is also known as the back-error propagation method.

Given a set of input–output pairs <X, D> and a neural network with more than one layers. The problem at hand is to find the weights of the neurodes for each layer. In the previous section, the delta rule has been presented for an SLP. The same notion can be extended to multiple layers. First, consider the output layer. Since each output neurode is independent of the others, the adaptation for each neurode in the output layer is the same. The update equation is obtained by repeatedly applying the chain rule to the cost function. Note that the update equation is applicable for an arbitrary activation function. The effects of the activation function are accounted for by the f'(net) term in the partial derivatives.

For neurodes in the hidden layer, the output of the neurodes is not directly given. However, with the use of the chain rule, the "desired" output can still be inferred. Begin with the squared error at the output layer again. In summary, the generalized delta rule can be extended to an arbitrary number of layers in the MLP. The activation function is arbitrary for any neurode in any layer. Likewise, the connectivity pattern can also be arbitrary. In other words, the network designed can trim the network and specifically allow or disallow connections to be made. Furthermore, there can also be fixed weights in the connectivity pattern. Those weights that are supposed to be fixed are simply never updated. Those connections that are not supposed to exist simply take on zero weight values. The generalized delta rule is able to work with an arbitrary architecture of the neural network and under a variety of constraints. Because of such flexibility, the generalized delta rule has been commonly used for many neural network applications.

Variations in the Back-Error Propagation Algorithm

Because of its importance and widespread acceptance, the generalized delta rule has been a subject of intense study. There are many variations to the basic generalized delta rule reported in the literature. Some of these variations are presented here.

One variation deals with the definition of the error. The original definition is called the single pattern error.

In this approach, the error is defined to be the error for each input pattern and to update the weight matrix after the presentation of each input pattern. This definition is straightforward but tends to be computationally intensive because all the weights have to be updated after each successive presentation of the input patterns. Due to the order of the presented patterns, it has been observed that sometimes the values of some of weights oscillate back and forth. Some patterns tend to pull the weights one way, while another pattern tends to pull the same weight in the opposite direction.

Since the oscillatory behavior partially comes from the order of the input patterns, one approach is to randomize the input pattern order. Define an epoch as a complete cycle of presenting all the input patterns once. The random approach would dictate that the order of presenting the patterns are random in each epoch. This approach tends to minimize the oscillatory behavior, thus speeding up the convergence rate in many cases. However, this approach is still computationally intensive since the weights are still updated upon each presentation of the input pattern.

In order to reduce the computational load, another approach is to eliminate the oscillatory behavior in the weight adaptation process. Hence, the weights are updated only after all the patterns are presented. Instead of using the error in one pattern and updating the weights immediately, all the patterns are presented first and the weights are updated according to the cumulative error.

The previous approach relies on the square of the error. This means that large errors tend to dominate the adaptation process. One proposal is to normalize the cumulative error so that the square root of the cumulative error is used.

Another approach is to simply use the absolute value of the error, commonly known as the L_1 norm instead of the Euclidean norm or the L_2 norm. For classification, it is the number of wrong classifications that is important. Hence, this is called the classification approach.

The actual deviations of the error values are of less importance. If the prediction is in error, then it is in error. How much in error is irrelevant.

Learning Rate and Momentum

The convergence rate of the update process is governed by the step size η or sometimes known as the learning constant. If η is small, then convergence is slow because the weights are updated by small increments. On the other hand, if η is large, the convergence is rapid because each update to the weight moves the weight a significant amount. However, if η is too large, then overshoot of the parameter values often occurs causing oscillatory behavior to occur. This leads to slow convergence again. In some cases, overshoots may also diverge.

From the standpoint of accuracy in estimation of the weight values, if η is small, then more accurate estimation is obtained because each update can only move the weight a small amount resulting in the weight not being able to wander around the target point. On the other hand, if η is large, the weight estimation is less accurate because the weight value can wander further from the ideal location.

The proper setting of the learning constant is very crucial. When the input data is known, it is possible to determine what the upper bound on η is. The optimal value must lie between 0 and the maximum value. Some researchers propose a learning schedule where the learning constant starts at a maximum value and gradually decreases in value as the iteration progresses.

In general, the convergence rate of the generalized delta rule is slow. To speed up the convergence, a momentum term is sometimes used. The first term on the right-hand side is the usual gradient term controlled by the step size η. The second term on the right-hand side is called momentum term and is dependent on the previous change. If the current change in the specific weight in question has the same sign as the previous change, then the momentum term enhances the change. This extra enhancement can exponentially increase when the requested change in the gradient is the same sign for consecutive steps.

The inclusion of the momentum term in the update equation makes the update itself an auto-regressive process. This process can be unraveled into a series representation. In other words, the momentum allows the gradient at a particular step to influence the updates for later steps. The use of the momentum generally increases the convergence rate of the adaptation. The amount of momentum to be added is controlled by a positive constant $\alpha>0$. However, care should be exercised in setting the values for α because too large a value may cause unnecessary oscillations in the update process.

There is a delicate balance between the choice of η and α because the two parameters are not independent from one another. In many application, even the momentum term is on a schedule.

Other Back-Error Propagation Issues

In the generalized delta rule, the algorithm has been shown to converge regardless of the initial values of the weights. However, the choice of the initial values of the weights does affect the convergence rate. It is obvious that if the values of the initial weights are close to the optimal values, the convergence would be rapid. On the other hand, if the initial weights are far from the optimal values, then the convergence depends on the value of the learning constant and the momentum constant.

While the back-error-propagation algorithm is able to determine the values of the weights in a neural network, the algorithm does not leave any hints as to the proper architecture of the neural network. The architecture of the neural network must be determined a priori.

When speaking about the architecture of a neural network, a user must determine the number of layers and the number of neurodes to use in each layer. In terms of the input layer, the input dimension is usually dictated by the application. Likewise for the output layer, the output dimension represents the number of categories or classifications required for the problem. Hence the output dimension is usually also dictated by the problem. As for the hidden layer, it is not easy to determine the appropriate number of neurodes. The general rule of thumb is to start with a large number and prune afterward or to start with a small number and slowly increase it.

Counter-Propagation Network

In previous sections, a single layer of competitive learning network can be used to automatically "discover" the clusters inherent in the input data. Quite often, it is highly desirable to label the classes or to combine the classes into a single class. This can be done by adding a special output layer after the competitive learning layer. An architecture of the combined network is shown.

There are two layers in the network. The first layer is the Kohonen network with competitive learning strategy. The purpose of this layer of the network is to automatically and adaptively locate the clusters in the input data space. The second layer is called the counter-propagation network or sometimes also called the outstar. This layer is also known as the Grossberg layer. The purpose of this network is primarily to combine clusters found in the first layer and to label the clusters in the desired output.

The training of the first layer has already been presented in earlier sections. Once the clusters are found, one of the neurodes in the Kohonen layer will be active upon presentation of a particular input sample data. Since the initial weights are random, which neurode will respond to the particular input sample is also random even though only one of the neurodes will be active. Hence the output of the first layer can be considered as a permutation vector where all elements of the vector are zero except one.

The addition of the second layer allows the network to manipulate the permutation vector. To combine clusters into a single class, a connection can be made from the output of the first layer to the input of the same neurode on the output layer. To "label" a particular class with a specific output pattern, simply use the weights of the second layer to generate the desired results. In other words, the weights of the second layer are trained according to the desired output. At this phase of the training, a sample pattern is presented to the first layer of the network. It is assumed that the Kohonen layer has already been trained. Therefore the weights are fixed. The presentation of a pattern causes one of the neurodes in the first layer to become active while all the outputs of all the rest of the neurodes are zero. The single activated neurode is connected to every neurode of the output layer in a star configuration. Hence it is called an outstar. Typically the activation function of the output layer is taken to be a linear ramp function. The weights of the activated neurode can now be trained. In fact, the weight is simply the desired output.

RADIAL BASIS NETWORKS

One of the primary uses of the back-error propagation is modeling. In other words, given samples of the input and output pairs, the neural network finds the proper transformation so that the input space can be properly and accurately mapped to the output space. The mapping can also be thought as function approximation. The transformation is the function that must be approximated from a set of given samples.

Function approximation can be accomplished in two ways: find a suitable function or interpolate. In the first method, the goal is identify a function and to estimate the parameters of that function so that the output of the function adequately produces the expected output values. Many techniques have been developed in this direction. Most techniques involve the user finding the form of the function, and the algorithm determines the best parameter values. Hence, the problem becomes a parameter estimation problem.

Another form of function approximation is interpolation. While interpolation is often less thought of as function approximation, but from another perspective, the interpolation function is in fact a function used for approximation. The former method is called parametric approximation because the function is fixed, and only the parameters are varied to fit the application. The later method is called nonparametric approximation because there is no specified form to the function, and the form of the function varies with the number of data points used.

Interpolation

To highlight the similarity of neural networks with interpolating functions, this section reviews some of the interpolation method, in particular the nearest neighbor interpolation, the Lagrange interpolation, and the spline interpolation.

In the nearest neighbor interpolation method, the closest neighbors to the unknown input point are selected. The function values of the unknown input point are then calculated in proportion to the selected points that are closest to the unknown point. The method works reasonably well and requires no training. However, the closest points must be determined before the interpolation formula is applied.

In the Lagrange interpolation, the interpolation is carried out by defining the Lagrange function. The form of the Lagrange function is predetermined and is formed from the given points. There is no need to select points for the interpolation as all sample points are used in the function.

In essence, the Lagrange function is the interpolating function used to approximate the model. What is so powerful about Lagrange interpolation is that the interpolating function changed according to the unknown input. This is in contrast to many function approximation methods including the MLP where one function is used to approximate the entire range of the input data.

In the Lagrange interpolation, the interpolated values at the given sample points are guaranteed to take on the given function values. The interpolating function is therefore continuous. This is also the case for the nearest neighbor approach. However, the derivatives are not continuous for the sample points. In the spline interpolation approach, the algorithm is designed in such a way that the derivatives at the sample points are also continuous.

Radial Basis Network

The derivation of the radial basis network is based on the principle of a regularizing network. While for a typical neural network, the goal is to minimize the cost function; here the cost function is usually taken to be the square error. For a regularizing network, and additional term, called the regularizing term is added. The object is to find a function (F) such that the cost function is minimized. The cost function is composed of two terms. The first term is the standard error term. The second term is the regularizing term based some operation (D) that is related to the derivative of the function sought. Note that when λ goes to zero, the cost function degenerates to the standard cost function. By including the regularizing term, the "smoothness" of the interpolating function can be controlled because the regularizing term is related to the derivatives of the function. Poggio (see Badiru & Cheung, 2002) has shown that the solution to minimizing such a cost function lies in the use of Green's functions. Radial basis function (RBF) network is a neural network composed of two layers. The first layer is the RBF layer, and the second layer is the encoding layer. Each neurode in the RBF layer is formed from a given sample point. The output of the neurode in the RBG layer is the Green's function.

The sample points act as a center. The unknown point is compared with the center, and the value of the neurode diminishes as the unknown point moves further away from the center, the sample point. This behavior is similar to the inverse of the Euclidean distance. In most cases, the Euclidean distance is radially symmetric, hence the name. There are a number of Green's functions proposed in the literature.

The neurodes in the second layer simply serve to implement the weights needed to relate all the RBF outputs together. This is similar to the interpolating function where the contribution of each of the centers is weighted. When compared with Lagrange Interpolation, the Green's function is comparable to the Lagrange function and the weights are the function values of the sample points.

Looking at the RBF from a slightly different perspective, the determination of the weights and other parameters of the RBFs is a parameter estimation problem. The only difference between a RBF network and an MLP is that we are now stipulating specific functional characteristics for the neurodes in the first layer. Instead of using the typical neurode (a linear combiner with a non-linear transformation) as the first layer, we are now using the RBF neurodes as the first layer. The second layer in both cases remains the same. For the MLP, the network parameters, namely the weights, are obtained by the generalized delta rule that stems from the successive application of the chain rule. While then functional form for the first layer may have been changed, the same procedure can still be applied.

So far, the RBF network is constructed with each RBF neurode corresponding to each given sample point. When the number of given sample points are large, a large RBF network naturally results. It is of interest of course to trim the network to a smaller size than the original. One way to trim the network is to use representative sample points instead of all the sample points. In other words, if clusters can be found a priori, then the location of the center or centroid of the clusters can be used instead of all the sample points in the cluster. This approach, however, adds an additional step in the analysis as the user must first determine the number and location of the clusters before applying the RBF network for modeling.

Another approach is to apply the generalized delta rule to the RBF network. For the first layer, the parameters to be estimated are the location (x_i) and the spread of the centers (C). The adaptation process now carries out the two tasks: finding the clusters and estimating the parameters of the clusters. For categorical data, the location of clusters provides an efficient way to summarize the data. For non-categorical data, a large number of clusters are required. For example, the approximation of a straight line will require a number of centers spaced throughout the line so that the approximation can be kept within the desired accuracy.

SINGLE-LAYER FEEDBACK NETWORK

In an earlier section, single feedforward neurode has been shown to simulate the operation of many combination circuit components. In digital logic, combinational circuits are formed entirely from feedforward circuits. Another major area of digital logic is sequential circuit. The basic building block in sequential circuit is a flip-flop. A flip-flop is composed with all combinational circuit components but is connected together with feedback. The feedback allows the flip-flop to "remember" previous information. Using flip-flops and other combinational circuit components, a sequential circuit is built that can remember information. A flip-flop stores a single bit of information. A shift register remembers a word.

If the output of a McCulloch–Pitts neurode is fed back as one of the inputs, then the neurode behaves like a flip-flop. Let X=[T, s, r, o], where o is the neuronal output fed back to itself after passing through a delay element, and let W=[], then the McCulloch–Pitts neurode behaves as an S-R flip-flop. In other words, when SR=00, the flip-flop output remains as is. If the previous output is 0, the next output is 0. If the previous output is 1, then the next output is 1. However, if the SR input is 10, then regardless of the previous output, the next output is 1. Likewise, if the SR input is 01, then regardless of the previous output, the next output is 0. This also implies that all sequential circuit elements can be replaced by McCulloch–Pitts neurodes, hence a neural network.

In a single-layer feedback network, the delayed output of each neurode is connected to the input of every other neurode except itself. In other words, there is no self-excitation.

The network is initialized by the input X producing an initial output. Once the network has been initialized, the network continues to update itself because the output is fed back to itself. The network will continue to change until the delayed output produces the exact same output. The network is then said to be in equilibrium.

A feedback network is called a dynamical system. For a feedforward network, the output is always just a combination of the inputs and does not change according to time as long as the input remains the same. For a feedback network, the input only initializes the network. Once initialized, the network output will continue to change. Depending on the system characteristics of the network, a dynamic system could continue to change or stabilize at an equilibrium point. In some cases, a dynamic system may diverge causing the output to grow unbounded. This usually happens when there is positive feedback. It is also possible for the system output to oscillate. This is called a limit cycle. The system neither converges nor diverges. The desirable case of course occurs when the dynamical system converges to equilibrium, i.e., a stable point.

For a dynamical system, there is a notion of an energy state related to the state of the network. The motion of a dynamical system is always toward the low energy state.

DISCRETE SINGLE-LAYER FEEDBACK NETWORK

A discrete feedback network is formed when the activation of the neurodes is discrete. The feedback network is shown with McCulloch–Pitts neurode. The network is initially started with the input i. This causes an initial output. The bias in the equation can be incorporated as a part of the input as before. After the initial presentation of the input, the pattern is removed and the output of the network is fed back to the input after a unit time delay. The energy level is a function of the current output of the neurodes and the weights. Hence, there is an energy level associated with the network at any time.

Since the network is dynamic, the training procedure is not so easy. Given a set of P input data sample points. In the training process, the idea is to adjust the weights so that given one of the input sample points as input to the network, the network produces the same output as the input. This is done with no self-excitation. In other words, there are no inconsistencies between the input and the output, a condition for stability. During the training process, the input sample is left at the input. The weights are then adjusted until the output equals the input. The weights are adjusted so that when the same pattern is presented at the input, there are no further inconsistencies in the network and the network gives the same pattern as output. At this point, the network is said to have been trained for that pattern. It is possible for a network to be trained to remember more than one input pattern. This can be achieved by repeatedly training each pattern in turn.

It can be shown that for a dynamic system, as the network changes, each change tends to cause the network to settle toward a lower energy state than before.

The above table shows that each change in the network output will always cause the energy to decrease or remain as is. In other words, the energy is non-increasing since the change in energy is always negative or zero.

During recall, the weights are fixed. When the network is initialized by an unknown input pattern, the initial output is fed back to the input and the network successively adjusted its output in response to changes in its own output. In summary, both the training and the recall for a single-layer feedback network are multistep. This is characteristic of a feedback network. Once the network has been trained, the trained samples can be retrieved.

When a new pattern is introduced at the output, the network immediately attempts to produce an output. If the input pattern is one of the trained patterns, the trained weights cause the network to produce an output that is consistent with the network. No further changes are produced. If the input pattern is not one of the trained patterns, then the network output is fed back to the input causing the network to change. Since every change of the network will cause the network state to go toward a low energy state, the network will eventually settled on one of the trained patterns as each of the trained pattern represents a low energy state. Hence, a discrete single-layer feedback network is sometimes called a content-addressable memory. This means that the stored memory can be retrieved by supplying part of the desired memory. To retrieve one of the stored patterns, only part of the stored patterns is required to initialize the network. Based on the partial input, the network proceeds to gravitate toward the lowest energy state of one of the stored patterns and thus regenerates in the process the complete stored pattern at its output.

Content-addressable memory can be used in a number of ways. One possible application is pattern recognition. A stored pattern contains both the pattern and the key for the class of the pattern. When an unknown pattern is presented to the network, the corresponding key according to the stored pattern is retrieved or regenerated by the network. In other words, the pattern has been recognized. Another application is image restoration. When a noisy pattern is presented to the network, the noisy pattern is gradually replaced by the stored pattern, thus "cleaning up" or restoring the image.

BIDIRECTIONAL ASSOCIATIVE MEMORY

A bidirectional associative memory (BAM) is a special case of a feedback network. Normally, the output of a neurode is fed back to the input of every neurode. In a BAM network, the single layer of neurodes is divided into two sections with the output of the neurodes from one section connected to all the neurode inputs in the other section and vice versa. In other words, instead of building the correlation through the weights between every input pixels with every other input pixels, the correlation here is built between the neurodes in one section with the neurodes from the other section.

As an associative memory, one section of the network can contain the data while the other section of the network contains the key of the associated data. When the data is presented to one section of the network, the key is regenerated by the other section of the network. Likewise, when the key is presented to the other side of the network, the data is regenerated by the key on the other side of the network. Likewise, the BAM network can also be used for pattern recognition. The network is initially trained with the patterns and the associated class label. When an unknown pattern is presented to the network, the key associated with the closest stored pattern is regenerated on the other section.

HOPFIELD NETWORK

Based on the properties of an electronic circuit arrangement, Hopfield proposed a feedback network. The Hopfield network is analogous to a single-layer feedback network. In Hopfield's original formulation, the network is composed of electronic components. A neurode is simulated by an operational amplifier. The input of the operational amplifier is a current sum. The currents are generated as a result of the all the outputs of the amplifiers through current-limiting resistors. These resistors are analogous to the weights of a neural network. Hence, the operational amplifier is in essence a linear combiner with input weights connected to the output of the other neurodes. Note that the input of the operational amplifier is also connected to a parallel resistive–capacitive network. The inclusion of the capacitor, a nonlinear component, provides the simulation of a nonlinear activation function.

The behavior of the circuit can be described according to the voltage and current relationship. With the components connected as shown, the nodal equation can be written at the input of the operational amplifier.

The nodal equation is called an equation of motion because it dictates how the neurode output, the voltage level, changes as a function of time. The energy function defined above has been shown to be the Lyapunov function for the network. Note that the Lyapunov energy function and the equation of motion go in pairs because the Lyapunov function is not unique.

One of the major applications of the Hopfield net is optimization. Hopfield originally proposed the use of the network to solve the traveling salesman person (TSP) problem. The TSP problem is an NP-complete problem. The problem can be stated as follows: Given the location of N cities, find the shortest path that connects all the cities and returning to the originating city. All the cities must be visited and visited once.

The solution to the TSP problem and likewise for any other optimization problems is first to define the objective function to be minimized. Since the Hopfield net tends to move toward a low energy state as defined by the Lyapunov function, the Lyapunov function can be used as the objective function. The solution is then found by applying the corresponding equation of motion to the network output. When the network converges to a low energy state, a possible solution to the optimization problem is found.

In solving the TSP problem using a Hopfield net, the first step is to find a way to represent the solution space. Given N cities to be visited, an array of NxN neurodes can be used. Each row represents a city to be visited, and each column represents the order of the route. If the output of the neurodes represents a permutation matrix with one 1 in each row and one 1 in each column, then the permutation matrix represents a possible and legitimate route. For example, given five cities (A, B, C, D, E), then a 5×5 array of neurodes is used. If the array output is [0 1 0 0 0; 1 0 0 0 0; 0 0 0 0 1; 0 0 1 0 0; 0 0 0 1 0], then the route is B→A→D→C→E→B.

The next step in solving the TSP problem is to define the objective function. Since the energy function has a quadratic form, the objective function must also be in a quadratic form. The primary objective of the TSP problem is to minimize the path length. Hence, the cost function is the path length.

In addition to the cost function, it is necessary to enforce further constraints to ensure that the solution is a legitimate route. Clearly if the sales person does not go anywhere, the path length would be zero and that is not an acceptable solution. Constraints are enforced by adding penalty to the cost function when the constraints are not satisfied. To enforce that each city is only visited once, each row of the permutation matrix must only contain a single one.

The final objective function is the weighted combination of the distance cost function and the three constraints formulated in quadratic forms.

Having found the desired objective function, the next step is to cast the objective function in the form of the Lyapunov function for the Hopfield net. Note that even though the formulation of the TSP problem indicates an array of neurodes and hence individually identified by a double index, in reality, the neurodes are in a single layer because the output of any neurode is fed back to the input of all other neurodes. Changing the energy function into a double index form and matching the objective function with the energy function, the weights of the neurodes can be found as follows.

Note that there is no training of the weights here. Rather, the weights themselves represent the optimization problem to be solved. In fact, the weights encapsulate the problem itself. So, the problem is solved during the "recall" mode. The network is initialized with random weights and then let loose. As a dynamic system, the network output is constantly changed while the weights are fixed. Each change causes the energy function to decrease, thus accomplishing the objective of minimizing the objective function. Eventually, the output settles in a low energy state representing a possible solution to the problem.

SUMMARY

A single neurode represents an elementary processing unit and can be used as a building block for a variety of systems and applications. The basic definition of a single neurode is a linear combiner followed by a nonlinear activation function. From such a simple processing element, very complex systems can be built. With proper choice of the input weights, a neurode could function like any Boolean components. Hence, as much as AND, OR, and NOT gates can be used to build powerful computers, collection of neurodes is also expected to be able to perform complex tasks.

Collection of neurodes into a single layer in a feedforward manner provides powerful mapping abilities such as associative memory, modeling, function approximation, and classification. The power of the neural network lies not only on the fact that the network is capable of performing the above tasks, but more so because the network is able to learn how to perform the tasks from given examples. In fact, the network is able to learn on its own.

When additional layers are cascaded together into a multiple layer network, it has been postulated that such a network is able to approximate an arbitrary function. The development of the generalized delta rule, or more commonly known as the back-error propagation method, further enhances the use of MLP. Complex and nonlinear models can now be modeled by the network.

When the output of the network is fed back to the input, a feedback network is obtained that functions like a dynamic system. A dynamic system is not only good for pattern recognition and image enhancement, but more importantly it can be used for solving optimization problems. By taking advantage of the dynamics of the network, a solution to the optimization problem can be iteratively found.

REFERENCES

Badiru, A. B., & J. Cheung (2002), *Fuzzy Engineering Expert Systems with Neural Network Applications*, John Wiley & Sons, New York.

Badiru, A. B., & D. B. Sieger (1998), "Neural network as a simulation metamodel in economic analysis of risky projects," *European Journal of Operational Research*, Vol. 105, pp. 130–142.

Milatovic, M., A. B. Badiru, & T. B. Trafalis (2000), "Taxonomical Analysis of Project Activity Networks Using Competitive Artificial Neural Networks," *Smart Engineering System Design: Neural Networks. Fuzzy Logic, Evolutionary Programming, Data Mining, and Complex Systems: Proceedings of ANNIE Conference*, ST. Louis, MO, Nov 5–8, 2000.

Sieger, D. B., & A. B. Badiru (1993), "An artificial neural network case study: prediction versus classification in a manufacturing application," *Computers and Industrial Engineering*, Vol. 25, Nos. 1–4, pp. 381–384.

Neural-Fuzzy Networks for Artificial Intelligence

5

TECHNOLOGY COMPARISONS

The field of computational intelligence or soft computing encompasses three main research directions: artificial neural networks, fuzzy logic, and evolutionary algorithms. Each area is well suited to different aspects of the problem-solving process. For the technical details on these, see Badiru and Cheung (2002), Badiru and Sieger (1998), Sieger and Badiru (1993), and Milatovic et al. (2000). The strength of neural networks lies in its ability to easily model unknown systems. One of the most popular neural network models is based on a nonlinear transformation of a linear combiner. Using the back-propagation algorithm, the network can be trained with input data to model an arbitrary system, i.e., to approximate an arbitrary function. Other neural network models such as counter-propagation networks and radial basis function (RBF) networks also provide function approximation using slightly different topology and training techniques. Furthermore, there is a whole set of other networks such as the Hopfield nets that are designed to solve open-ended optimization problems. A third type of neural networks such as Kohonen's maps can be used to discover clustering through a self-organizing weights update algorithm. With a slightly different training algorithm, a single-layer perceptron with added lateral connections can also be configured to do principal components analysis (PCA). PCA is another form of representing the input data with only the salient features with minimal dimensions.

DOI: 10.1201/9781003089643-5

While neural networks are ideal for modeling known or unknown associations that exist between the input and output data, significant data cleaning and preprocessing are usually needed. In other words, input data must be carefully coded and prepared for the network to process. Another difficulty with neural networks is that the network must first be trained. The more input data there is, the better are the training results. The richer the input data there is, the more accurate is the model. However, training requires substantial amount of time and resources. These difficulties restrict the widespread use of neural networks in many applications. In many decision-making systems, it is important to be able to explain the process that the decision is made. It is not a simple matter to derive rules from neural networks.

The main concept for fuzzy logic is to use unsharp boundaries of membership functions to describe the implicitly imprecise concepts in data representation. From this perspective, fuzzy logic is ideally suited for user interactions and data representation. Since fuzzy logic is also numerical in nature, concepts can be expressed and manipulated as mathematical variables. Using the extension principle, most of the crisp operations can be readily adapted to fuzzy operations. In the crisp domain, models are made using regression or autoregressive-moving average representations. Likewise in the fuzzy domain, fuzzy models can also be made using fuzzy regression and fuzzy operators. Hence, fuzzy operations include both logical operations and numerical operations. Another useful feature of fuzzy logic is its ability to make inferences. Propositions are readily represented by fuzzy values. Since implication is also a fuzzy operator, approximate reasoning can be carried out naturally as fuzzy computations.

The concepts of fuzzy logic clearly complement those in neural networks. While fuzzy logic provides simple data representation, neural network provides none. Where fuzzy logic can be used to model a system, neural networks are well suited to provide sophisticated models for diverse types of systems. However, if there is prior knowledge about the underlying system, fuzzy logic can readily encapsulate the knowledge in terms of rules and relations, while it is not particularly easy to preprogram a neural network with prior knowledge. Given a set of training samples, it is not simple to train a fuzzy model, but many algorithms have been developed in the past for training neural networks.

Another aspect of computational intelligence is evolutionary algorithm. This type of algorithm is biologically inspired. The principle idea is that a solution can be produced through genetic reproductions among a population of viable individuals, each individual representing a possible solution. There are two main classes of evolutionary algorithms: genetic algorithm and evolutionary programming. Genetic algorithm uses genes, collectively called a chromosome as the basis to represent possible solutions. Solutions are paired using crossover operations to produce new solutions. Mutations are used to enrich the genetic pool of the population and to explore unchartered territories of the search space. Evolutionary programming places less emphasis on the genetic structure and uses mutation as the primary operation to reproduce offspring.

Evolutionary programming is a search methodology and is suited for solving open-ended optimization problems. While neural networks have been shown to solve open-ended problems such as the traveling salesperson problem, detailed analysis shows that the neural network spends significant amount of time converging to a local minimum. Simulated annealing has been proposed as a method to cause the neural network to settle on a global minimum, but the technique is computational intensive because the annealing temperature must be lowered very slowly. Evolutionary programming provides a significantly more efficient way to search since each step of the algorithm produces a whole new generation of solutions. In neural networks, the objective function and all subsequent constraints must be explicitly programmed into the weights. In evolutionary algorithms, the algorithm is independent of the objective functions and associated constraints. The algorithm only requires that there is a cost function associated with each solution.

Comparing evolutionary programming with fuzzy logic, the two technologies are complementary. In fuzzy logic, open-ended search can be obtained through forward or backward chaining performed in an orderly fashion. Quite often the search is exhaustive, hence the technique is good for problems with a small solution set. In evolutionary algorithm, the solution is obtained by randomly generating individual solution; hence the technique is ideally suited to problems with a large solution set.

For complex systems, no single technology can easily satisfy all the requirements of the problem. It is natural to combine more than one technology in the quest for the solution to the problem on hand. These are called hybrid systems. Hybrid systems are designed to take advantage of the strengths of each system and to avoid the limitations of each system also. For example, it is natural for neural networks to learn, but it is cumbersome for a fuzzy system to learn. Hence a combination of the two would result in a rule-based system that can learn and adapt. On the other hand, learning in neural network is slow. Hence there are many proposed hybrid systems where a fuzzy system is used to tune the learning rate and momentum terms in an effort to speed up the convergence rate. In systems where prior knowledge is available, what is known can be easily coded in rules and facts, but it is not a simple matter to encode prior knowledge in a neural network. These are only small sample of examples where hybrid systems would be ideal. In this chapter, we examine the synergism between neural networks and fuzzy logic.

NEURONS PERFORMING FUZZY OPERATIONS

One of the simplest types of hybrid systems is to train a neural network to perform fuzzy logic operations. The main advantage of this is the reduction of time complexity. Since there are many neural network chips available through many vendors, these chips are capable of executing billions of neural connections a second. If a neural network can be trained to perform fuzzy operations, then the fuzzy operations can also be performed with the same speed. This is a definite advantage to emulating the fuzzy operations using microcontrollers or simulating the operations using computer instructions.

Three most basic operations in classical set theory are AND, OR, and NOT operations. The corresponding operations in fuzzy logic are min, max, and complement. When a fuzzy variable can only take on the two extreme values (0 and 1), then the fuzzy operations degenerate into the respective classical set operations. In a more general setting, the terms conjunctive, disjunctive, and complement are used to represent intersection, union, and complement operations.

NEURONS EMULATING FUZZY OPERATIONS

A simple neuron can be made to perform logic functions with some special arrangement. This section presents the neural network arrangement for conjunctive, disjunctive, and complement networks.

A conjunctive network performs the intersection operation for fuzzy variables. Using the definition of standard t-norm, the intersection is the minimum operation on all fuzzy inputs. A standard feed-forward neural network with special input arrangement can be used to perform this operation. The network is designed to find the minimum of the input fuzzy values. It is assumed that the crisp inputs have already fuzzified. Assume that the fuzzified values are p_i, $i=1...N$. In order for the network to work, the fuzzy inputs are first ordered giving p'_i, $i=1...N$. Then the difference from consecutive inputs is obtained. For the conjunctive network, the weighting function is predefined to be 1/n. The hard-limiter output is a one whenever the argument is greater than or equal to 1, otherwise the hard-limiter output is a 0. The activation function is simply taken to be the linear function. Hence, the output of the neuron is simply the weighted sum of the input with the connection weights.

Using the same basic architecture, a disjunctive network can also be obtained. A disjunctive network performs the union operation for fuzzy variables. Using the definition of standard t-conorm, the union is the maximum operation on all fuzzy inputs. A standard feed-forward neural network with special input arrangement can also be used to perform this operation.

The network is designed to find the maximum of the input fuzzy values. It is assumed that the crisp inputs have already fuzzified. Assume that the fuzzified values are p_i, $i=1...N$. In order for the network to work, the fuzzy inputs are first ordered. For the disjunctive network, the weighting function is predefined to be all 1. Define the connection weights v_i. The hard-limiter output is a 1 whenever the argument is greater than or equal to 1, otherwise the hard-limiter output is a 0. The activation function is simply taken to be the linear function. Hence, the output of the neuron is simply the weighted sum of the input with the connection weights.

In like manner, a complement network can also be designed. A complement operation is a unary operation, in that the complement is applied only to one fuzzy variable. Using the definition of the standard complement operation, the output is simply the difference of the fuzzy input from one. Hence there are two inputs in the complement network: the first is the fuzzy variable and the second is a constant one. The corresponding connection weights for the two inputs are –1 and 1.

A hybrid neuron is also a neuron with crisp inputs and crisp outputs. However, instead of performing a weighted sum followed by a nonlinear transformation, a fuzzy neuron performs one of the fuzzy operations such as the t-norm or the t-conorm operation. While this adaptation of a neuron may not be biologically based, the topology and architecture are certainly biologically inspired.

Corresponding to the crisp inputs, a hybrid neuron also has crisp weights. In general, arithmetic operations such as multiplication and addition are not used in combining inputs and weights because these functions tend to produce resultant values that do not necessarily lie in the interval between 0 and 1. Instead, fuzzy operations are preferred so that the resultant values do lie in the interval between 0 and 1. Each input and its corresponding weight can be combined using a continuous operation such as t-norm or t-conorm. The aggregation of all weighted inputs can also be performed with any of the fuzzy continuous operations. If a nonlinear transformation is required, a continuous function mapping the aggregation value to the output is used.

A hybrid AND neuron takes on two crisp inputs and produces a single crisp output. Corresponding to each input is a crisp connection weight. Each input and its associated weight are combined using a disjunctive (union) operation (C(x, y)). The weighted inputs are then aggregated together by a conjunctive (intersection) operation (T(x, y)). Using C to denote the t-conorm and T the t-norm operations, the output of the hybrid AND neuron can be denoted. Likewise, a hybrid OR neuron takes on two crisp inputs and produces a single crisp output. Corresponding to each input is also a crisp connection weight. Each input and its weight are combined using a conjunctive (intersection) operation (T(x, y)). The weighted inputs are then aggregated together by a disjunctive (union) operation (C(x, y)). The output of the hybrid OR neuron can be denoted. A hybrid AND neuron and a hybrid OR neuron can be diagrammed.

NEURAL NETWORK PERFORMING FUZZY INFERENCE

One of the strengths of a fuzzy logic system is its ability to make inferences. Model characteristics are usually written in facts and rules. An example of a rule is:

If x is X_i and y is Y_i then z is Z_i If x is X_i and y is Y_i then z is Z_i

The rule stated that if the input variable x belongs to the membership X_i and the input variable y belongs to the membership Y_i, then the output variable z would belong to Z_i. There are a number of approaches in realizing a rule and a rule set. These approaches are explored in this section.

REGULAR NEURAL NETWORK WITH CRISP INPUT AND OUTPUT

The if–then rule points to a system having two crisp inputs and a single crisp output. Though the rule deals with fuzzy variables, x, y, and x are crisp by themselves. The fuzzy value describes the degree that the crisp value belongs in the X_i, Y_i, and Z_i membership functions, respectively. Note that the output is also a crisp value. It is now easy to see that the if–then rule can be viewed as a black box with two crisp inputs and one crisp output. As such, the rule can be modeled by a regular neural network such as a multilayer perceptron.

If the membership functions of the input and output variables are known a priori, then values of the membership functions can be sampled and used as a training set for the neural network, i.e., $<(x, y), z>$ where the first set of values in the double are input parameters and the second parameter of the doublet is the output. If training samples are used to training the if–then rule, then the same training samples can likewise be used also to train the neural network. The mapping from the if–then rule to the neural network is direct and straightforward. If there are more inputs and/or outputs, then the corresponding neural network would also have the same number of inputs and outputs.

REGULAR NEURAL NETWORK WITH FUZZY INPUT AND OUTPUT

For some problems, the input may not be a crisp value, but rather a fuzzy value defined by the associated membership functions. A regular neural network can still be used in this case. One approach is to sample the membership function with discrete number of domain values. Instead of working with a continuous interval, the membership function is sampled at discrete values. In this case, the input to the neural network is a set of membership values at discrete locations of the input parameter. The shape of the membership curve is represented by the function values at the selected locations. Likewise, the output membership curve is also represented by a series of function values at discrete points. Representation in this format is very powerful because a rule can now be formed for the entire membership function.

If X and Y then Z

Here, X, Y, and Z are membership functions. Using a series of crisp values, each of the membership functions can be sampled. The entire series for X and that for Y serve as inputs to the neural network. Likewise, the entire series serve as the output from the neural network. The training sequence is then a double $<(x_1, x_2,\ldots, x_n; y_1, y_2,\ldots, y_n), (z_1, z_2,\ldots, z_n)>$ where the first parameter contains the two sequences for X and Y, and the second parameter contains the sequence for the output Z. The neural network can now be repeatedly trained using the standard back-propagation method or other standard techniques until the network output yields the desired result.

Uehara and Fuhise have proposed a variation to this scheme. Instead of discretizing the domain and sampling the membership functions at those discrete points, they proposed that the membership function be represented by a series of α-cuts. Each α-cut represents an interval. In this case, the input to the neural network would be a series of interval values for different α-cut values. Likewise, the output from the neural network would also be a series of interval values for the same corresponding α-cut values. Regardless of the discretization approaches used, the membership function can easily be reconstructed.

FUZZY INFERENCE NETWORK

With some careful rearranging of the inputs, a single neuron has been shown to function as a fuzzy AND, fuzzy OR, and a fuzzy complement operator. But the real power lies in the ability of a neural network to emulate the fuzzy inference process.

In approximate reasoning, the system is represented by a set of rules and facts. Facts are inputs obtained from the system environment. Rules describe the model characteristics. Using a set of predefined membership functions, crisp inputs are converted to fuzzy variables. Rules relate the fuzzy input variables to fuzzy output variables. The antecedent of each rule is constructed with conjunction and disjunction of the fuzzy input variables. The inference is made using the implication operator based on the generalized modus ponens, modus tollens, and hypothetical syllogism. After the inference has been carried out, the fuzzy output variables are defuzzified to yield a crisp number for output.

The process described above can be emulated by a neural network with a topology similar to a multilayer perceptron. Ideally, the input to the network is crisp numbers and the output of the network is also crisp numbers. The first layer of the neural network is to fuzzify the input values. The fuzzification process can be performed by a layer of RBF neurons or by special subnetworks designed to emulate the membership functions. Each RBF neuron emulates a single membership function, hence a set of RBF neurons is required to produce an array of fuzzy values for each crisp input.

The second layer of the neural network implements the conjunctive and/or disjunctive operation of the fuzzy inputs for the antecedents of each fuzzy rule. Neurons that can emulate conjunction and disjunction have been presented in the previous section. These specialized neurons are used to combine the fuzzy input variables produced by the first layer. If the rule antecedents are overly complex, the operations can be emulated by more than one layer here. One suggestion is a conjunctive layer followed by a disjunctive layer.

The third layer of the neural network implements the implication operation. The fuzzy value of the antecedent is used to limit the degree of truthfulness of the output membership functions. This also is a conjunctive operation. To realize the output membership functions, a set of neurons is used to represent the possible crisp output possibilities. The output of the neurons in this layer is limited by the truthfulness of the antecedents obtained from the previous layer.

The fourth layer is the consequent layer. This is a disjunctive layer as the consequent is usually taken to be the t-conorm from all the inferences. The final conclusion is the cumulated truthfulness, i.e., the union from all the inferences.

The last layer is the defuzzification process. A single neuron is used for each crisp output. The weights are arranged in such a way to emulate one of the defuzzification methods. The most common one is the centroid method where the crisp output is the centroid of the fuzzy output values.

ADAPTIVE NEURO-FUZZY INFERENCE SYSTEM (ANFIS)

An adaptive neuro-fuzzy inference system or adaptive network-based fuzzy inference system (ANFIS) is a kind of artificial neural network that is based on adaptive inferencing framework. Through dual integration, it has the potential to capture the benefits of both neural networks and fuzzy logic in a single framework. Its inference system corresponds to a set of fuzzy IF–THEN rules that have learning capability to approximate nonlinear functions. Hence, ANFIS is considered to be a universal estimator. For using the ANFIS in a more efficient and optimal way, one can use the best parameters obtained by genetic algorithm. The bulk of the network is used for fuzzification using two layers of perceptrons. The inference is based on sigma-pi neurons, and the output membership function is not used. The output of the network directly yields a crisp value.

The first layer consists of two-input perceptrons with the usual sigmoidal transformation. This layer produces a series of sigmoidal curves. The first input comes from the crisp input value. The second input is always one representing the bias. A bias is needed to offset the crisp input values. This has the same effect as shifting the sigmoidal function to be centered at the desired value. The neurons at this layer are similar to other neural networks using a single layer of perceptrons.

A second layer is next used to collate the sigmoidal outputs together to form membership functions. This layer is composed of two-input neurons with linear activation function. The connection weights are simply 1 and –1. Typically, two sigmoidal functions are needed to make one membership function. A membership function is realized by taking the difference between two sigmoidal function outputs.

The next layer is composed of a set of sigma-pi neurons. The output of these neurons is the product of the weighted inputs instead of the sum of the weighted inputs. The fuzzy inputs are weighted and multiplied to realize the conjunctive relation of the antecedent. In this case, the intersection of the various fuzzy inputs is realized using the product rule instead of the standard (min) rule.

The last layer is simply a set of regular neurons with linear activation functions. The output of each neuron in this layer is the weighted sum of the previous layer. The output of the previous layers represents the strength of activation of a particular rule. This layer collates the strengths of different rule activations together to produce a set of crisp output values.

Another variation to the same network is to eliminate the second layer entirely. If a Gaussian transformation is used as the activation function instead of the sigmoidal function, then the shape of the Gaussian function can be used to represent the membership function. If there are a lot of linguistic variables, this can greatly simplify the network topology.

The key point to observe here is that by emulating the fuzzy rules in neural network architecture, the network can now be trained with standard back-propagation methods in response to training patterns. This means that the shape of the membership functions and the strength of the connection for the rules can be adjusted and learned. When the training is completed, the neural network can simply be converted back to fuzzy rules if desired. This is the primary advantage of using a neural network to emulate fuzzy inference.

COMMUTATIVE APPLICATIONS

A direct application of emulating fuzzy logic using a neural network is to use the output of the network as a way to tune parameters of another neural network. It has often been observed that the learning rate and momentum greatly affect the ability of a neural network to converge. However, there is no simple way to select the proper values of the learning rate and the momentum factor.

It has also been suggested that an adaptive scheme could possibly be more effective during the training phase of the process.

This approach has been proposed by Hertz and Hu (1992). Hertz and Hu used a second neural network to adaptively adjust the learning rate according to a set of heuristic rules. The second neural network emulates the heuristic rules and produces the recommended values for the learning rate to be used in the first network during its training process. To accomplish this, a neural network is used with one input and one output parameter. The crisp input parameter is the error for the present iteration. The crisp input value is first fuzzified into seven membership functions (NL, NM, NS, ZE, PS, PM, PL) corresponding to whether the error is positive and negative and whether the magnitude of the error is great or small. Hertz and Hu developed a number of heuristic rules, which are preprogrammed into the neural network. The collation of all rules yields four the fuzzy values for four membership functions (ZE, PS, PM, PL) for the learning rate. The output of the network is a single variable, the learning rate. The four fuzzy output values are defuzzified into a single crisp value.

Another application was proposed by Baglio et al. (1994) in modeling urban traffic noise. The goal is to predict the degree of urban noise from passage of motor vehicles. However, the noise is often mollified due to the shadowing effects of buildings and building elevation. Their approach compared the use of a traditional neural network against the use of a fuzzy inference network. It was found that the performance of the fuzzy inference network is comparable to the traditional neural network. However, the fuzzy inference network has a significant lower computational complexity than a traditional network.

CLUSTERING AND CLASSIFICATION

In clustering, while the training patterns are given, the exact grouping of these patterns is unknown (Sieger & Badiru, 1993). By repeatedly seeing the training patterns, a neural net or fuzzy system is used to categorize the patterns into distinct groups or clusters. In pattern classification, not only are the training samples given, but also the cluster that each training sample belongs to is also known. The task of the neural network or the fuzzy system is to learn the association so that the system can successfully recognize the input patterns according to the proper class.

If a neural network is used for the recognition process, then the input to the neural net is a doublet <$(x_1, x_2, \ldots, x_n)$, Class>, where the first parameter represents the input pattern and the second parameter represents the class that the pattern belongs in. Since a neural network can accommodate large input dimension, it is not unusual that the input set consists of the original raw data set. If a fuzzy logic system is used, then the doublet can immediately be written in the form of a rule.

If x_1 is X_1 and x_2 is X_2 and...and x_n is X_n then Class.

When rules are used to describe an input pattern, rules are more efficient when a small number of antecedents are used. Hence, the raw data pattern is often preprocessed to reduce the dimension of the original data set. A number of transformations can be used including Fourier transform for one-dimensional signal and two-dimensional images. Other transforms used are PCA and singular value decompositions. If more reduction in data dimension is required, then a feature extraction process is usually performed first. The input to the rules would then be extracted features. In many cases, these features are meaningful to human recognition. Hence the reasoning from a fuzzy logic system can be explained in more recognizable terms to the user.

Classification

Consider a simple two-class system with two inputs. Each dimension of the input space is divided by a set of membership functions. The boundaries of the membership functions separate the input space into distinct areas. Each area can now be labeled with the class number. By further subdividing the input parameters into additional membership functions, it is clear that any clusters can be formed on the input space.

Assume that the input parameters have been set up each with two input parameters, i.e., $x_1 \in$ {Small, Large} and $x_2 \in$ {Small, Large}. Then there are four areas in the input space. This corresponds to four rules.

If x_1 is Small and x_2 is Small then Class One.

If x_1 is Small and x_2 is Large then Class One.

If x_1 is Large and x_2 is Small then Class Two.

If x_1 is Large and x_2 is Large then Class Two.

While the number of membership functions is known and set a priori, the exact shape and location of the membership functions can be varied for accurate results. This is when training samples are used to adapt the shape and locations of the membership functions.

The same problem can be cast into a neural network paradigm proposed by Sun and Jang. The neural network would have two input parameters and a single output parameter.

The inputs to the neural network would be the two input parameters, both crisp. The first layer of the neural network is to compute the degree of likeness between the crisp input parameter and the membership function. These neurons can be RBF neurons or regular neurons with Gaussian activation functions. For the Gaussian activation function, the adjustable parameters are the width and centroid of the Gaussian curve. Other membership functions including triangular and trapezoidal shapes can also be used. For triangular membership function, the adjustable parameters are the left lower limit, the center upper limit, and the right lower limit. For trapezoidal membership function, the adjustable parameters are the left lower limit, upper limit, right upper limit, and the right lower limit. Regardless of the activation function used, each parameter of the activation function can be adapted by the standard back-propagation approach.

The second layer of the neural network emulates the conjunctive operation. This is used to connect the various parts of the antecedent together. The output of this layer is the strength of the present rule indicating how closely the input parameters match the stated membership functions. If the match is good, then the strength of the rule is strong; however if the match is poor, then the strength of the rule is greatly diminished.

The third layer of the neural network is a single-layer perceptron. The output of each neuron is the weighted sum of the firing strengths of the rules in previous layer. The weighted sum is subjected to a nonlinear activation function that is normally taken to be the sigmoidal transformation. In the present case, since there are only two classes, a hard limiter can be used to indicate whether the output is Class One or Class Two.

During training, the training data consists of the input data and its associated class identification. Since the third layer is a single-layer perceptron, the weights can be adapted by the standard back-propagation approach. The second layer implements the conjunctive operation. Most of the time, there are no adjustable parameters in this layer. For the first layer of the neural network, the parameters there can also be adjusted by the same concept of the back-propagation approach.

MULTILAYER FUZZY PERCEPTRON

Nauck and Kruse (see Badiru and Cheung, 2002) developed a special three-layer perceptron called fuzzy perceptron. The system is called NEFCLASS that stands for a neuro-fuzzy system for the classification of data developed.

The fuzzy perceptron is designed to learn from training samples the separation for the different classes. The knowledge of the pattern classification is contained in a set of fuzzy rules.

The fuzzy perceptron is composed of three layers, the input layer, the hidden or rule layer, and the output layer. The first layer performs the fuzzification of the crisp input parameters. The second layer implements the antecedents of the rule set. The last layer performs the defuzzification. Layer 1 performs the fuzzification process. Each neuron in this layer inputs one crisp parameter and outputs a series of fuzzy values according to the set of membership functions defined for that parameter. Each output fuzzy value indicates the degree of match between the input crisp value and the associated linguistic concept. Layer 2 performs the conjunctive operation for selected fuzzy values. Each neuron in this layer implements one fuzzy rule. The output of the neuron is the activation strength of the associated rule and is obtained by a fuzzy AND of the various fuzzy membership values. Layer 3 performs the defuzzification process by combining the activation strengths of all the rules together to form an estimate of the class. It should be noted that the output does not actually yield an estimate of a specific class, rather the output shows an estimated possibility of each class. If desired, a fourth layer such as a MAXNET can be used to interpret the results by selecting the class with the largest output.

To start, the user must define the basic structure of the fuzzy perceptron. It is necessary to define the number of neurons in the hidden layer and an initial estimate of the various membership functions in the input layer. Alternatively, the neurons in the hidden layer can be added iteratively during training. When an input pattern is submitted to the network, a search is performed to see what set of input fuzzy values would yield the best output. The set of selected fuzzy values is then inserted into the hidden layer if there are no other neurons in the hidden layer representing the same set of input fuzzy values. If the system is small enough, then it is possible to start with all possible combinations. After training, a scoring method is used to gauge the effectiveness of each rule and any poor performing neurons are then trimmed from the system.

REFERENCES

Badiru, A. B., & Cheung, J. (2002), *Fuzzy Engineering Expert Systems with Neural Network Applications*, John Wiley & Sons, New York.

Badiru, A. B., & D. B. Sieger (1998), "Neural network as a simulation metamodel in economic analysis of risky projects," *European Journal of Operational Research*, Vol. 105, pp. 130–142.

Baglio, S., L. Fortuna, M. G. Xibilia, & P. Zuccarini (1994), "Neuro-fuzzy to predict urban traffic," in *Proceedings of EUFIT94 Conference*, Aachen, Germany, 1994, pp. 20–29.

Hertz, D., & Q. Hu (1992), "Fuzzy-neuro controller for backpropagation networks," in *Proceedings of the Simulation Technology and Workshop on Neural Networks Conference*, Houston, TX, 1992, pp. 540–574.

Milatovic, M., A. B. Badiru, & T. B. Trafalis (2000), "Taxonomical Analysis of Project Activity Networks Using Competitive Artificial Neural Networks," *Smart Engineering System Design: Neural Networks. Fuzzy Logic, Evolutionary Programming, Data Mining, and Complex Systems: Proceedings of ANNIE Conference*, ST. Louis, MO, Nov 5–8, 2000.

Sieger, D. B., & A. B. Badiru (1993), "An artificial neural network case study: prediction versus classification in a manufacturing application," *Computers and Industrial Engineering*, Vol. 25, Nos. 1–4, pp. 381–384.

Index

academia–industry cooperation for expert systems 30
adaptive correlation matrix memory 74
adaptive neuro-fuzzy inference system 103
AI branches 12
AI software 19
ANFIS 103
application roadmap 27
associative memory 71

back-error propagation 79
back-error propagation algorithm 80
benefits of expert systems 24
bidirectional associative memory 89
branches of artificial intelligence 12

classification 106
clustering and classification 105
clustering by Hebbian learning 76
clustering by Oja's normalization 77
collaboration, digital 57
commutative applications 104
competitive learning network 78
correlation matrix memory 72
counter-propagation network 83

data processing 25
data processing to knowledge processing 25
DAU SE model 50
Defense Acquisitions University SE Model 50
DEJI systems model 45
digital collaborations 57
digital data input–process–output 53
digital engineering and systems engineering 44
digital framework for AI 43
digital systems framework for AI 43
discrete single-layer feedback network 88

emergence of expert systems 15
error-correcting pseudo-inverse method 74
evolution of smart programs 9
expert systems 15, 19
expert systems: the software side of AI 19
expert systems characteristics 19
expert systems structure 21
expert systems, the need for 23

feedforward network 78
first AI conference 8
first conference on artificial intelligence 8
future directions for expert systems 29
fuzzy inference network 102

heuristic reasoning 25
historical background of AI 2
hopfield network 90
human intelligence 4
humans in the loop 61

industry partnership 30
interpolation 84
introduction to AI 1

knowledge processing 25
lean 60

lean and six sigma in AI 60
learning network 78
learning rate and momentum 81
LMS approach 73

machine intelligence 4
McCullough-Pitts Neurode 69
multi-category SLP 71
multilayer fuzzy perceptron 107
multiple-layer feedforward network 78
multiple-layer perceptron 78

natural language dichotomies 6
neural network 13
neural network performing fuzzy inference 100
neural network with crisp input and output 100
neural networks 13
neural networks for artificial intelligence 65
neural-fuzzy networks 95

neurode 67
neurons emulating fuzzy operations 98
neurons performing fuzzy operations 98
no-news 7
no-salt 7

object-oriented analysis and design (OOAD) 52
Oja's normalization 77
OOAD 52
origin of artificial intelligence 2

perceptron 78
principal components 75
propagation 79

radial basis networks 84
regular neural network 101

sample of expert systems applications 33
self-organizing network 74
single neurode as binary classifier 70
single neurode perceptron 71
single-layer feedback network 87
single-layer feedforward network 71
six sigma 60
smart programs 9
sodium free 7
spiral model 49
symbolic processing 27

technology comparisons 95
transition from data processing 25

user interface 26

V-model 48

walking skeleton model 50
waterfall model 47
Widrow-Hoff approach 73

XOR Example 79